Information Security and Cryptography

For further volumes:
www.springer.com/series/4752

Benny Applebaum

Cryptography
in Constant Parallel Time

 Springer

Benny Applebaum
School of Electrical Engineering
Tel Aviv University
Tel Aviv, Israel

ISSN 1619-7100   Information Security and Cryptography
ISBN 978-3-662-50713-1   ISBN 978-3-642-17367-7 (eBook)
DOI 10.1007/978-3-642-17367-7
Springer Heidelberg New York Dordrecht London

To my parents, my wife and my children.

Foreword

This book provides excellent coverage of the exciting results obtained in the last decade regarding the complexity of doing cryptography.

Cryptography is concerned with the study of the design of systems that are easy to operate but hard to abuse. Thus, a complexity gap between the ease of using such systems and the difficulty of abusing them lies at the heart of cryptography. The question addressed in the current book is how wide can this gap be. In a nutshell, the work presented in this book asserts that the gap may be much wider than one would have thought: The systems may be extremely easy to use (i.e., each output bit can be computed based on *a constant number* of input bits), whereas no efficient procedure may abuse them (i.e., the notion of security is the standard one).

Let me be somewhat more technical. The work provides strong evidence that many cryptographic primitives and tasks can be implemented with very low complexity. For example, it shows that the existence of one-way functions that can be evaluated in NC^1 (and even somewhat above NC^1) implies the existence of one-way functions that can be evaluated in NC^0. Whereas the former are widely believed to exist (e.g., based on the standard factoring assumption), most researchers have previously believed that the latter do not exist. Recall that evaluation in NC^0 means that each output bit only depends on a constant number of input bits. This work further shows that dependence on *four* input bits suffices (whereas dependence on at least *three* input bits is definitely necessary).

Let me briefly discuss the aforementioned beliefs. Recall that all known constructions of cryptographic primitives are based on complexity assumptions. In particular, all these assumptions (and actually also the very existence of these cryptographic primitives) imply $P \neq NP$ and thus establishing any of these assumptions would resolve the famous P-vs-NP question. Thus, unless one resolves the P-vs-NP question, a result of the current type must be based on some assumptions. The complexity assumptions used in the current work are among the weakest ones used in cryptographic research.

Actually, the work presents a transformation of implementations of cryptographic primitives, taking any implementation in a class between NC^1 and NC^2, and producing an implementation in NC^0. (Recall that NC is the class of problems that are solvable by polynomial-size circuits of polylogarithmic depth, and NC^i denotes the subclass in which the exponent of the polylogarithmic function is i.) The transformation is based on "randomizing polynomials" a notion introduced a few years ago for very different purposes. In particular, the original motivation was the study of *information-theoretic* privacy in multi-party computation, whereas the current context is complexity theoretic in nature.

The centerpiece of the book is presented in Chaps. 3 and 4, where the aforementioned results are proved. Let me stress that these chapters present an amazing breakthrough in the study of the theoretical foundations of cryptography. In particular, they provide extremely efficient implementations of several basic cryptographic tools. As I noted above, this outstanding achievement took almost all experts in the area by surprise. The following chapters (i.e., Chaps. 5–8) provide intriguing follow-ups and extensions of the direction initiated in Chaps. 3 and 4. The book also contains a nice exposition of the relevant background (specifically, Chap. 2).

Weizmann Institute of Science   Oded Goldreich
May 2013

Preface

Cryptography is concerned with communication and computation in the presence of adversaries. A fundamental challenge in theoretical and practical cryptography is to minimize the computational complexity of honest parties while providing security against computationally strong attackers. Ideally, one would like to construct cryptographic tools or "primitives" which can be computed extremely fast and retain strong security guarantees. These two targets, *efficiency* and *security*, are somewhat contradictory as highly efficient functions may be too simple to generate cryptographic hardness. Identifying the minimal level of efficiency which still guarantees security is therefore a major research goal.

This book studies this question through the lens of parallel-time complexity. We ask whether basic cryptographic primitives can be computed in constant parallel time. Formally, we consider the possibility of computing instances of these primitives using $\mathbf{NC}^0$ circuits, in which each output bit depends on a constant number of input bits. Despite previous efforts in this direction, there has been no convincing theoretical evidence supporting this possibility, which was posed as an open question in several previous works (e.g., [50, 69, 85, 105, 112]). We essentially settle this question by providing strong evidence for the possibility of cryptography in $\mathbf{NC}^0$. In particular, we derive the following results.

Existence of Cryptographic Primitives in $\mathbf{NC}^0$ We show that many cryptographic primitives can be realized in $\mathbf{NC}^0$ under standard intractability assumptions used in cryptography, such as those related to factoring, discrete logarithm, or lattice problems. This includes one-way functions, pseudorandom generators, symmetric and public-key encryption schemes, digital signatures, message authentication schemes, commitment schemes, collision-resistant hash functions and zero-knowledge proofs. Moreover, we provide a *compiler* that transforms an implementation of a cryptographic primitive in a relatively "high" complexity class into an $\mathbf{NC}^0$ implementation. This compiler is also used to derive new unconditional $\mathbf{NC}^0$ *reductions* between different cryptographic primitives. In some cases, no parallel reductions of this type were previously known, even in $\mathbf{NC}$. Interestingly, we get *non-black-box* reductions.

Pseudorandom Generators with Linear Stretch in NC^0   The aforementioned constructions of pseudorandom generators (PRGs) were limited to stretching a seed of n bits to $n + o(n)$ bits. This leaves open the existence of a PRG with a linear (let alone superlinear) stretch in NC^0. We construct a linear-stretch PRG in NC^0 under a relatively new intractability assumption presented by Alekhnovich [5]. We also identify a new connection between such pseudorandom generators and hardness of approximations for combinatorial optimization problems. In particular, we show that an NC^0 pseudorandom generator with linear stretch implies that Max 3SAT cannot be efficiently approximated to within some multiplicative constant. Our argument is quite simple and does not rely on PCP machinery.

Cryptography with Constant Input Locality   After studying NC^0 functions, in which each output bit depends on a constant number of input bits, we move on to study functions in which each *input* bit affects a constant number of output bits, i.e., functions with constant *input* locality. We characterize what cryptographic tasks can be performed with constant input locality. On the negative side, we show that primitives that require some form of non-malleability (such as digital signatures, message authentication, or non-malleable encryption) *cannot* be realized with constant input locality. On the positive side, assuming the intractability of certain problems from the domain of error correcting codes, we obtain new constructions of one-way functions, pseudorandom generators, commitments, and semantically secure public-key encryption schemes whose input locality is constant. Moreover, these constructions also enjoy constant *output locality*. Therefore, they give rise to cryptographic hardware that has constant-depth, constant fan-in and constant *fan-out*.

A Study of Randomizing Polynomials   Most of our results make use of the machinery of *randomizing polynomials*, which were introduced by Ishai and Kushilevitz [92] in the context of information-theoretic secure multiparty computation. Randomizing polynomials allow us to represent a function $f(x)$ by a low-degree randomized mapping $\hat{f}(x, r)$ whose output distribution on an input x is a *randomized encoding* of $f(x)$. We present several variants of this notion along with new constructions. Our new variants have applications not only in the domain of parallel cryptography. For example, by extending the notion of randomizing polynomials to the computational setting, we show that, assuming a PRG in NC^1, the task of computing an *arbitrary* (polynomial-time computable) function with computational security can be reduced to the task of securely computing degree-3 polynomials (say, over $\mathbb{F}_2$) without further interaction. This gives rise to new, conceptually simpler, constant-round protocols for general functions.

This Version   This book is based on the author's doctoral dissertation which was submitted to the Technion in 2007. Some of the sections and proofs have been extended to provide more details and intuition. The content has also been updated to reflect the main recent developments in the field of parallel-time cryptography. A detailed chapter-by-chapter description of the contents and a high-level list of updates appear in Sects. 1.2.2 and 1.3.

Tel Aviv, Israel   Benny Applebaum

Acknowledgements

I am greatly indebted to Yuval Ishai and Eyal Kushilevitz, my advisors, for their guidance and friendship. Over the years, I have spent many hours in conversations with Yuval and Eyal. These long discussions have taught me invaluable lessons regarding many aspects of the scientific work and have shaped my scientific outlook. For the enjoyable collaboration which led to this book, for their insightful and knowledgeable advice, and for all their patience and help, I am deeply grateful to Yuval and Eyal.

I am greatly indebted to Oded Goldreich for closely accompanying this research. I was fortunate to have Oded as the editor of the journal versions of some of the works that appear in this book, and was even more fortunate to collaborate with him on some follow-up works. Oded's numerous suggestions and comments have significantly improved this monograph in many ways, and I am most grateful to him for sharing his wisdom and insights with me.

During my graduate studies, I had the opportunity to discuss research topics with many friends and colleagues. These interactions have been pleasant and fruitful. For this, I would like to thank Omer Barkol, Rotem Bennet, Eli Ben-Sasson, Eli Biham, Iftach Haitner, Danny Harnik, Moni Naor, Erez Petrank, Omer Reingold, Ronny Roth, Amir Shpilka, Amnon Ta-Shma, Enav Weinreb and Emanuele Viola. I am also thankful to Eli Biham, Oded Goldreich, Erez Petrank, and Omer Reingold for serving on my thesis committee.

I would like to thank all the people from the Computer Science department in the Technion, with whom I have worked and studied, for making my time at the Technion so pleasant. Special thanks go to my office partner, Boris Kapchits, and my floor mates, Rotem Bennet, Niv Buchbinder, Oren Katzengold, Jonathan Naor and Sharon Shoham—I really liked all these endless coffee breaks!

Since my graduation, I have spent wonderful years in the CS departments of Princeton University and the Weizmann Institute of Science. My sincere thanks to the theory groups at these institutes, and especially to Boaz Barak, Oded Goldreich, Moni Naor and Avi Wigderson for memorable times and priceless lessons from which I have learned so much.

I would like to acknowledge my current academic home, Tel Aviv University's School of Electrical Engineering. Special thanks go to my close colleagues Guy Even, Boaz Patt-Shamir and Dana Ron, for their kind support and lovely company.

Finally, I am grateful to my family and friends for their love and support. Most significantly, I would like to thank my parents, Arie and Elka, and my wife, Hilla. Some feelings cannot be expressed with words, but I can sincerely say that none of this would have been possible without you!

Contents

Chapter 1
Introduction

Abstract This book studies the parallel-time complexity of cryptography. Specifically, we study the possibility of implementing cryptographic tasks by NC^0 functions, namely by functions in which each output bit depends on a constant number of input bits. We provide strong evidence for the possibility of cryptography in NC^0, settling a longstanding open question.

1.1 The Basic Question

Cryptography is concerned with communication and computation in the presence of adversaries. In the last few decades, the theory of cryptography has been extensively developed and has successfully provided solutions to many security challenges. Moreover, due to the evolution of the Internet, cryptographic tools are now widely employed both by individuals and by organizations. Daily actions, such as checking an account balance or electronic commerce, make essential use of cryptographic primitives such as encryption schemes and digital signatures.

In this book, we study the computational cost of cryptography. For concreteness, let us focus on two basic cryptographic tasks: computing a function which is hard to invert (i.e., *one-way function*), and computing a function which expands a short random seed into a longer "random-looking" string (i.e., *pseudorandom generator* [36, 144]). Our goal is to understand the minimal computational complexity of these cryptographic primitives. Pushing this question to an extreme, it is natural to ask whether such primitives can be made "computationally simple" to the extent that each bit of their output is only influenced by a constant number of input bits, independently of the desired level of security. Specifically, the following fundamental question was posed in several previous works (e.g., [50, 69, 85, 105, 112]):

> Is it possible to compute one-way functions, or even pseudorandom generators so that every bit of the output can be computed by reading a constant number of bits of the input?

The class of functions which are computationally simple in the above sense is denoted by NC^0. We let NC^0_c denote the class of NC^0 functions in which each of the output bits depends on at most c input bits, and refer to the constant c as the *output locality* of the function (or locality for short).

The above question is qualitatively interesting as it explores the possibility of obtaining cryptographic hardness using "extremely simple" functions. However, func-

B. Applebaum, *Cryptography in Constant Parallel Time*,
Information Security and Cryptography,
DOI 10.1007/978-3-642-17367-7_1, © Springer-Verlag Berlin Heidelberg 2014

tions in $\mathbf{NC}^0$ might be considered to be too degenerate to perform any interesting computational tasks, let alone cryptographic tasks which are perceived to be inherently complex. Indeed, all common implementations of cryptographic primitives not only require each output bit to depend on many inputs bits, but also involve rather complex manipulations of these bits.

The possibility of cryptography in $\mathbf{NC}^0$ has been studied since the mid-1980s. Several works made progress in related directions [50, 66, 69, 85, 91, 108, 112, 116, 117, 138, 147], conjecturing either the existence (e.g., [69]) or the non-existence (e.g., [50]) of cryptographic primitives in $\mathbf{NC}^0$. However, despite all this body of work there has been no significant theoretical evidence supporting either a positive or a negative conclusion.

In this book, we provide a surprising affirmative answer to this question. We prove that most cryptographic primitives can be implemented by $\mathbf{NC}^0$ functions under standard intractability assumptions commonly used in cryptography (e.g., that factoring large integers is computationally hard). Specifically, primitives such as one-way functions, encryption, digital signatures, and others can be computed by extremely "simple" functions, in which every bit of the output depends on only four bits of the input.

This result is of both theoretical and practical interest. From a theoretical point of view, it is part of a general endeavor to identify the minimal resources required for carrying out natural computational tasks. From a more practical point of view, an $\mathbf{NC}^0$ implementation of a cryptographic primitive supports an ultimate level of parallelism: in an $\mathbf{NC}^0$ function, different output bits can be computed in parallel without requiring intermediate sequential computations. In particular, such functions can be computed in *constant* parallel time, i.e., by constant-depth circuits with bounded fan-in. Thus, $\mathbf{NC}^0$ primitives may give rise to super-fast cryptographic hardware.

1.1.1 Further Perspectives

It is instructive to examine the question of cryptography in $\mathbf{NC}^0$ from three distinct perspectives:

- **Applied cryptography**. In the community of applied cryptography it is widely accepted that functions with low locality should not be used as block ciphers or hash functions. Indeed, several central design principles of block ciphers (e.g., so-called Confusion-Diffusion [130], Avalanche Criterion [61], Completeness [98] and Strict Avalanche Criterion [141]) explicitly state that the input-output dependencies of a block cipher should be complex. In particular, in his seminal paper Feistel asserts that: "The important fact is that all output digits have potentially become very involved functions of all input digits" [61]. (In fact, this concern dates back to Shannon [130].) It is easy to justify this principle in the context of block-ciphers (which are theoretically modeled as pseudorandom functions or permutations), but it is not clear whether it is also necessary in other cryptographic applications (e.g., one-way functions, pseudorandom generators, or probabilistic public-key encryption schemes).

- **Complexity theory**. The possibility of cryptography in $\mathbf{NC}^0$ is closely related to the intractability of Constraint Satisfaction Problems. Inverting a function in $\mathbf{NC}_c^0$ can be formulated as a Constraint Satisfaction Problem in which each constraint involves at most c variables (c-CSP). For example, finding an inverse of a string $y \in \{0, 1\}^n$ under a function $f : \{0, 1\}^n \to \{0, 1\}^n$ is equivalent to solving a CSP problem over n variables $x = x_1, \ldots, x_n$ of the form:

$$\begin{cases} y_1 = f_1(x), \\ \vdots \\ y_n = f_n(x), \end{cases} \tag{1.1}$$

  where f_i is the function that computes the i-th output bit of f and y_i is the i-th bit of y. If f is in, say, $\mathbf{NC}_4^0$ we get an instance of a 4-CSP problem. Constraint satisfaction problems are well studied in complexity theory and are known to be "hard" in several aspects. In particular, the Cook-Levin theorem [48, 106] shows that it is NP-hard to exactly solve 3-CSP problems, while the PCP theorem [21, 22] shows that it is NP-hard even to find an approximate solution. It should be noted that, for several reasons, NP-hardness does not imply cryptographic hardness. Hence, although these results might indicate that it is not easy to invert $\mathbf{NC}^0$ functions in the worst case, they fall short of proving the existence of one-way functions in $\mathbf{NC}^0$.
- **Theoretical cryptography**. In modern cryptography it is typically assumed that honest parties are computationally weaker than malicious adversaries. Specifically, honest parties are limited to fixed polynomial time, while adversaries may have superpolynomial-time resources. Hence, it is crucial to assume the existence of a *computational gap* between the cost of computing a primitive (e.g., computing a one-way function) and the cost of breaking its security (e.g., inverting a one-way function). This computational miracle ("hard problems can be easily generated") lies at the heart of modern cryptography. The possibility of cryptography in $\mathbf{NC}^0$ suggests that the gap between efficiency and security can be much larger, as even extremely simple local functions can generate cryptographic hardness.

1.2 Our Research

Our main goal is to draw the theoretical and practical limitations of cryptography in constant parallel time. Hence, we try to characterize the precise computational power needed to fulfill different cryptographic tasks. In particular, we will be interested in questions of the form: Can a cryptographic primitive $\mathscr{P}$ be realized in some low complexity class WEAK? If so, what are the minimal assumptions required for such an implementation? We usually instantiate these meta-questions with the class $\mathbf{NC}^0$, but we will also consider other complexity classes, such as sub-classes of $\mathbf{NC}^0$ in which the output locality is bounded by some specific constant (e.g., $\mathbf{NC}_3^0$), or the class of functions whose *input locality* is constant (i.e., functions in which each bit of the input affects a constant number of output bits).

1.2.1 Our Approach

Our key observation is that instead of computing a given "cryptographic" function f, it might suffice to compute a related function $\hat{f}$ which (1) preserves the cryptographic properties of f; and (2) admits an efficient implementation. To this end, we rely on the machinery of *randomized encoding*, which was introduced in [92] under the algebraic framework of *randomizing polynomials*. A randomized encoding of a function $f(x)$ is a randomized mapping $\hat{f}(x, r)$ whose output distribution depends only on the output of f. Specifically, it is required that: (1) there exists a decoder algorithm that recovers $f(x)$ from $\hat{f}(x, r)$, and (2) there exists a simulator algorithm that given $f(x)$ samples from the distribution $\hat{f}(x, r)$ induced by a uniform choice of r. That is, the distribution $\hat{f}(x, r)$ hides all information about x except for the value $f(x)$.

We show that the security of most cryptographic primitives is inherited by their randomized encoding. This gives rise to the following paradigm. Suppose that we want to construct some cryptographic primitive $\mathscr{P}$ in some low complexity class WEAK. Then, we can try to encode functions from a higher complexity class STRONG using functions from WEAK. Now, if we have an implementation f of the primitive $\mathscr{P}$ in STRONG, we can replace f by its encoding $\hat{f} \in$ WEAK and obtain a low-complexity implementation of $\mathscr{P}$. This approach is extensively used in this book.

Non-black-Box Techniques In order to encode a function f, we apply a "compiler" to the *description* of f (given in some computational model). This technique is inherently *non-black-box* and, in some cases, it also yields parallel non-black-box transformations between different primitives. That is, the "code" of the NC^0-reduction we get, implementing a primitive $\mathscr{P}$ using an oracle to a primitive $\mathscr{Q}$, depends on the code of the underlying primitive $\mathscr{Q}$. This should be contrasted with most known transformations in cryptography, which make a black-box use of the underlying primitive. We believe that our work provides further evidence for the usefulness of non-black-box techniques in cryptography.

1.2.2 Results and Organization

In the following we give an outline of our results. Some of these results are also summarized graphically in Tables 1.1, 1.2, 1.3 and Fig. 1.1. The material presented in this book was obtained in joint works with Yuval Ishai and Eyal Kushilevitz [13–17].

Chapter 2—Preliminaries and Definitions We set the basic notation and definitions used in this book. This includes the notions of statistical and computational indistinguishability and some of their properties, as well as definitions and conventions regarding several computational models. We also define the main complexity classes mentioned in this book and investigate some of their simple properties.

Chapter 3—Randomized Encoding of Functions   We define the main variants of randomized encoding, including several *information-theoretic* variants as well as a *computational* variant. We investigate several useful properties of this notion, and discuss some of its limitations. Most of the material in this chapter is based on Sect. 4 of [15].

Chapter 4—Cryptography in $\mathbf{NC}^0$ We show that randomized encoding preserves the security of many cryptographic primitives. We also construct an (information-theoretic) encoding in $\mathbf{NC}_4^0$ for any function in $\mathbf{NC}^1$ or even in $\oplus\mathbf{L}/poly$. (The latter complexity class contains, in particular, functions computable in logarithmic space.) This result is obtained by relying on the constructions of [93] which give a low degree encoding for these classes. The combination of these results gives a compiler that takes as an input a code of an $\mathbf{NC}^1$ or log-space implementation of some cryptographic primitive and generates an $\mathbf{NC}_4^0$ implementation of the same primitive. This works for many cryptographic primitives such as OWFs, PRGs, one-way permutations, trapdoor-permutations, collision-resistant hash functions, encryption schemes, message authentication codes, digital signatures, commitments and zero-knowledge proofs. The existence of many of the above primitives in $\mathbf{NC}^1$ is a relatively mild assumption, implied by most number-theoretic or algebraic intractability assumptions commonly used in cryptography. We remark that in the case of two-party primitives (e.g., encryption schemes, signatures, commitments, zero-knowledge proofs) our transformation results in an $\mathbf{NC}^0$ sender (i.e., the encrypting party, committing party, signer or prover) but does not promise anything regarding the parallel complexity of the receiver (the decrypting party or verifier).[1] In fact, we prove that, in all these cases, the receiver *cannot* be implemented by an $\mathbf{NC}^0$ function, regardless of the complexity of the sender. (See Table 1.1.) Our techniques can also be applied to obtain unconditional constructions of non-cryptographic PRGs. In particular, building on [112], we obtain an ε-biased generator in $\mathbf{NC}_3^0$, answering an open question posed in [112]. The material in this chapter is mainly based on [15].

Chapter 5—Computationally Private Randomizing Polynomials and Their Applications   We consider a relaxed notion of randomized encodings, where the "hiding" property of randomized encoding is relaxed to the computational setting. We construct such an encoding in $\mathbf{NC}_4^0$ for every *polynomial-time* computable function, assuming the existence of a OWF in $\oplus\mathbf{L}/poly$ or $\mathbf{NL}/poly$. We present several applications of computationally private randomized encoding. In particular, we considerably relax the sufficient assumptions for $\mathbf{NC}^0$ constructions of cryptographic primitives (see Table 1.1), obtain new *unconditional* $\mathbf{NC}^0$ transformations between primitives (see Fig. 1.1), and simplify the design of constant-round protocols for multiparty computation. This chapter is based on [14].

---

[1] An interesting feature of the case of commitment is that we can also improve the parallel complexity at the receiver's end. Specifically, it can be implemented by an $\mathbf{AC}^0$ circuit (or even by a weaker circuit family). This feature of commitment carries on to some applications of commitments such as distributed coin-flipping and ZK proofs.

Table 1.1 Sufficient conditions for $\mathbf{NC}^0$ implementations of different primitives. In the case of PRGs (resp. collision-resistant hashing) we get an $\mathbf{NC}^0$ implementation with *sublinear* stretch (resp. shrinkage). **General assumptions**: We write "∃ in $\mathscr{C}$" to denote the assumption that the primitive can be realized in the complexity class $\mathscr{C}$. When $\mathscr{C}$ is omitted, we refer to the class $\mathbf{P}$, that is we assume that the primitive can be realized at all. The complexity classes $\mathbf{NL}/poly$ and $\oplus\mathbf{L}/poly$ are (incomparable) variants of uniform log-space computation. (Both classes contain the classes $\mathbf{L}/poly$ and $\mathbf{NC}^1$ and are contained in $\mathbf{NC}^2$. See Sect. 2.3 and Fig. 2.2.) We write "EOWF" (Easy OWF) to denote the existence of a OWF in either $\mathbf{NL}/poly$ or $\oplus\mathbf{L}/poly$. The symbol "×" denotes an impossibility result. **Concrete assumptions**: We use DLOG, LRSA, and DDH to denote the intractability of the discrete logarithm problem, the RSA problem with low exponent, and the decisional Diffie-Hellman problem, respectively. For example, the public-key encryption entry states that we can get a scheme in which the encrypting party is in $\mathbf{NC}^0$ assuming EOWF and the existence of a public-key encryption scheme. Moreover, these assumptions are implied by the intractability of either the factoring problem, the DDH problem or lattice problems. This entry also says that there is no such scheme in which the decryption is realized by an $\mathbf{NC}^0$ function

Primitive	General assumption	Concrete assumption
One-way function	∃ in $\oplus\mathbf{L}/poly$, $\mathbf{NL}/poly$	factoring, DLOG, lattices
One-way permutation	∃ in $\oplus\mathbf{L}/poly$	LRSA, DLOG
Trapdoor permutation	∃ in $\oplus\mathbf{L}/poly$	LRSA
Pseudorandom generator	∃ in $\oplus\mathbf{L}/poly$	factoring, DLOG, lattices
Collision-resistant hashing	∃ in $\oplus\mathbf{L}/poly$	factoring, DLOG, lattices
Public-key encryption		
Encrypting	∃ + EOWF	factoring, DDH, lattices
Decrypting	×	
Symmetric encryption		
Encrypting	EOWF	factoring, DLOG, lattices
Decrypting	×	
Signatures, MACs		
Signing	EOWF	factoring, DLOG, lattices
Verifying	×	
Non-interactive commitment	∃ + EOWF	factoring, DLOG, lattices
2-round stat. hiding commitment		
Committing	∃ in $\oplus\mathbf{L}/poly$, $\mathbf{NL}/poly$	factoring, DLOG, lattices
Verifying	×	
NIZK for $\mathbf{NP}$		
Proving	∃ + EOWF	factoring
Verifying	×	
Constant-round ZK proof for $\mathbf{NP}$		
Proving	∃ + EOWF	factoring, DLOG, lattices
Verifying	×	

Fig. 1.1 New NC^0 reductions. *Doubleline arrows* denote non-black-box reductions. We write "min-PRG" to denote a pseudorandom generator G with minimal stretch, i.e., $G : \{0, 1\}^n \to \{0, 1\}^{n+1}$

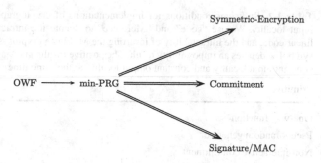

Chapter 6—One-Way Functions with Optimal Output Locality In Chap. 4 it is shown that, under relatively mild assumptions, there exist one-way functions (OWFs) in NC_4^0. This result is not far from optimal as there is no OWF in NC_2^0. In this chapter we partially close this gap by providing an evidence for the existence of OWF in NC_3^0. This is done in two steps: First, we describe a new variant of randomized encoding that allows us to obtain a OWF in NC_3^0 from a "robust" OWF which remains hard to invert even when some information on the preimage x is leaked; Second, we show how to construct a robust OWF assuming that a random function of (arbitrarily large) constant locality is one-way. The latter assumption is closely related to a conjecture made by Goldreich [69]. This chapter is based on [13].

Chapter 7—On Pseudorandom Generators with Linear Stretch in NC^0 The constructions of PRGs in NC^0 from the previous chapters were limited to stretching a seed of n bits to $n + o(n)$ bits. This leaves open the existence of a PRG with a linear (let alone superlinear) stretch in NC^0. We construct a linear-stretch PRG in NC_4^0 under a relatively new intractability assumption presented by Alekhnovich [5]. The linear stretch of this PRG is essentially optimal as there is no PRG with superlinear stretch in NC_4^0 [112]. (See Table 1.3.) We also show that the existence of a linear-stretch PRG in NC^0 implies non-trivial hardness of approximation results *without relying on PCP machinery*. In particular, it implies (via a simple proof) that Max3SAT is hard to approximate to within some multiplicative constant. The material of this chapter is based on [16].

Chapter 8—Cryptography with Constant Input Locality We study the possibility of carrying out cryptographic tasks by functions in which each input bit affects a constant number of output bits, i.e., functions with constant *input* locality. (Our previous results have only addressed the case of a constant *output* locality, which does not imply a constant *input* locality.) We characterize what cryptographic tasks can be performed with constant input locality. On the negative side, we show that primitives that require some form of non-malleability (such as digital signatures, message authentication, or non-malleable encryption) *cannot* be realized with constant (or, in some cases, even logarithmic) input locality. On the positive side, assuming the intractability of some problems from the domain of error correcting

Table 1.2 Sufficient conditions for implementations of cryptographic primitives with constant input locality. We use "code" and "McEliece" to denote the intractability of decoding random linear code, and the intractability of inverting the McEliece cryptosystem [110], respectively. The symbol × denotes an impossibility result. The positive results allow an implementation that enjoys constant input locality and constant output locality at the same time

Primitive	Assumption
One-way function	code
Pseudorandom generator	code
Non-interactive commitment	code
Public-key encryption	McEliece
Symmetric encryption	code
Signatures, MACs	×
Non-malleable encryption	×

Table 1.3 Pseudorandom generators: stretch vs. locality. A PRG has sublinear (resp. linear) stretch if it stretches n bits to $n + o(n)$ bits (resp. $n + \Omega(n)$ bits). A parameter is marked as optimal ($\checkmark$) if when fixing the other parameters it cannot be improved. The first construction requires the existence of a PRG in $\oplus \mathbf{L}/poly$, while the other two are based on concrete assumptions

Stretch	Output locality	Algebraic degree	Input locality	Reference
Sublinear	4	3	–	Theorem 4.6
Linear $\checkmark$	4	3	–	Theorem 7.3
Sublinear	3 $\checkmark$	2 $\checkmark$	3 $\checkmark$	Theorem 8.3

codes, we obtain new constructions of OWFs, PRGs, commitments, and semantically secure public-key encryption schemes whose input locality is constant. (See Table 1.2.) Moreover, these constructions also enjoy constant *output locality*. Therefore, they give rise to cryptographic hardware that has constant-depth, constant fan-in and constant *fan-out*. As a byproduct, we also construct a pseudorandom generator whose output and input locality are both optimal, namely, 3. (See Table 1.3.) Our positive results rely on a new construction of randomized encoding with constant input locality, while the negative results shed some light on the limitation of such an encoding. This chapter is based on [17].

1.3 Comparison with Doctoral Thesis Submitted to Technion

This book is based on the author's doctoral dissertation that was submitted to the Technion in June 2007. Except for stylistic changes, most of our presentation closely follows the original text. Notable exceptions include Chap. 6 which was significantly revised, and Chap. 8 which was revised and extended based on the journal version of [14]. Also, the order of the chapters was slightly changed: Chap. 6, which originally appeared as the last chapter of the dissertation, was moved to its current place.

Apart from these changes, we have added references throughout the book to some of the recent works in the area of parallel cryptography. The main developments are summarized below.

From OWF to PRG in Parallel Haitner, Reingold and Vadhan [83], strengthening the work of Håstad et al. [86], obtained an NC^1 transformation from one-way functions to pseudorandom generators. This result, combined with the techniques of this book, allows us to relax the sufficient conditions required for NC^0 implementations of complex primitives. For example, instead of requiring a log-space computable PRG, it suffices to require a log-space computable OWF. Similarly, this gives rise to new NC^0 transformations from one-way functions to complex primitives improving previous transformations that relied on minimal-stretch PRGs. The relevant chapters of this book (Chaps. 4 and 5) have been updated accordingly.

Cryptographic Hardness of Random Local Functions Random NC^0 functions are obtained by connecting each output to a random set of d inputs, and computing the output by evaluating some public d-local predicate over the corresponding inputs. The cryptographic hardness of such functions, which was first conjectured by Goldreich [69] and was employed in Chaps. 6 and 7, was extensively studied in the last few years. Several works have further investigated the security of random NC^0 functions as one-way functions [38, 39, 47] and as pseudorandom generators [11, 12]. Moreover, assuming the one-wayness of random local functions, it was shown how to construct public-key encryption schemes [11], NC^0 PRGs with large stretch [10], and NC^0 universal one-way hash function with linear shrinkage [20]. It is unknown how to construct the latter two primitives via the randomized encoding based approach. For more details, see Sects. 6.5 and 7.6.

Cryptography with Spatial Locality A new notion of *spatial locality* was introduced and studied in [18]. Roughly speaking, a circuit has a constant spatial locality if it can be embedded in the physical space (e.g., on a two-dimensional grid) such that the distance traveled by signals from an input node to an output node is bounded by an absolute constant. Spatial locality is strictly stronger than both constant input locality and constant output locality. In [18] several positive and negative results are obtained regarding the possibility of spatially local cryptography. The positive results have been further used to argue that for some simple dynamical systems (modeled as cellular automata) computational intractability arises from *almost all* initial configurations even after a *single* step of computation. See Sect. 8.8 for details.

Finally, we mention that locally computable cryptography has found further applications in the areas of hardness of approximation [10], lower-bounds for computational learning theory [11], leakage-resilient cryptography [94], and cryptography with optimal sequential-time complexity [95].

Chapter 2
Preliminaries and Definitions

Abstract This chapter presents the basic notation and definitions used in this book. This includes the notions of statistical and computational indistinguishability and some of their properties (Sect. 2.2), as well as definitions and conventions regarding several computational models (Sect. 2.3). We also define the main complexity measures mentioned in this book, namely, output locality, input locality, and algebraic degree, and we investigate some of their simple properties (Sect. 2.4).

2.1 General

Basic Notation By default, all logarithms are to the base 2. Let $\mathbb{N}$ denote the set of positive integers. For a positive integer $n \in \mathbb{N}$, denote by $[n]$ the set $\{1, \ldots, n\}$. For a string $x \in \{0, 1\}^*$, let $|x|$ denote the length of x. For a string $x \in \{0, 1\}^n$ and an integer $i \in [n]$, let x_i denote the i-th bit of x. Similarly, for $S \subseteq [n]$, let x_S denote the restriction of x to the indices in S. We will write $x_{\oplus i}$ to denote the string x with the i-th bit flipped. For a prime p, let $\mathbb{F}_p$ denote the finite field of p elements. Let $\mathbb{F}$ denote an arbitrary finite field. We will sometimes abuse notation and identify binary strings with vectors over $\mathbb{F}_2$. All vectors will be regarded by default as column vectors. Let $\langle \cdot, \cdot \rangle$ denote inner product over $\mathbb{F}_2$, i.e., for $x, y \in \mathbb{F}_2^n$, $\langle x, y \rangle = \sum_{i=1}^{n} x_i \cdot y_i$ where arithmetic is over $\mathbb{F}_2$. For a function $f : X \to Y$ and an element $y \in Y$, let $f^{-1}(y)$ denote the set $\{x \in X | f(x) = y\}$. Let $\text{Im}(f)$ denote the set $\{y \in Y | f^{-1}(y) \neq \emptyset\}$. A function $\varepsilon(\cdot)$ from positive integers to reals is said to be *negligible* if $\varepsilon(n) < n^{-c}$ for any $c > 0$ and sufficiently large n. We will sometimes use neg($\cdot$) to denote an unspecified negligible function.

Probabilistic Notation Let U_n denote a random variable that is uniformly distributed over $\{0, 1\}^n$. Different occurrences of U_n in the same statement refer to the same random variable (rather than independent ones). If X is a probability distribution, we write $x \leftarrow X$ to indicate that x is a sample taken from X. Let support(X) denote the *support* of X (i.e., the set of all elements with non-zero probability), and let $\mathbb{E}(X)$ denote the expectation of X. The *min-entropy* of a random variable X is defined as $H_\infty(X) \stackrel{\text{def}}{=} \min_x \log(\frac{1}{\Pr[X=x]})$. Let $H_2(\cdot)$ denote the binary entropy function, i.e., for $0 < p < 1$, $H_2(p) \stackrel{\text{def}}{=} -p \log(p) - (1 - p) \log(1 - p)$.

B. Applebaum, *Cryptography in Constant Parallel Time*,
Information Security and Cryptography,
DOI 10.1007/978-3-642-17367-7_2, © Springer-Verlag Berlin Heidelberg 2014

Adversarial Model By default we refer to an efficient adversary as a family of polynomial-sized circuits, or equivalently to a probabilistic polynomial-time algorithm that on input of size n gets an advice string of size $\text{poly}(n)$. However, all of our results also apply in a uniform setting in which adversaries are probabilistic polynomial-time algorithms.

2.2 Statistical and Computational Indistinguishability

The *statistical distance* between discrete probability distributions X and Y is defined as $\|X - Y\| \overset{\text{def}}{=} \frac{1}{2}\sum_z |\Pr[X = z] - \Pr[Y = z]|$. Equivalently, the statistical distance between X and Y may be defined as the maximum, over all boolean functions T, of the *distinguishing advantage* $|\Pr[T(X) = 1] - \Pr[T(Y) = 1]|$. For two distribution ensembles $X = \{X_n\}$ and $Y = \{Y_n\}$, we write $X \equiv Y$ if X_n and Y_n are identically distributed, and $X \overset{s}{\equiv} Y$ if the two ensembles are *statistically indistinguishable*; namely, $\|X_n - Y_n\|$ is negligible in n.

A weaker notion of closeness between distributions is that of *computational* indistinguishability: We write $\{X_n\}_{n\in\mathbb{N}} \overset{c}{\equiv}_{\delta(n)} \{Y_n\}_{n\in\mathbb{N}}$ if for every (non-uniform) polynomial-size circuit family $\{A_n\}$, the distinguishing advantage $|\Pr[A_n(X_n) = 1] - \Pr[A_n(Y_n) = 1]|$ is bounded by $\delta(n)$ for sufficiently large n. When the distinguishing advantage $\delta(n)$ is negligible, we write $\{X_n\}_{n\in\mathbb{N}} \overset{c}{\equiv} \{Y_n\}_{n\in\mathbb{N}}$. (We will sometimes simplify notation and write $X_n \overset{c}{\equiv} Y_n$.) By definition, $X_n \equiv Y_n$ implies that $X_n \overset{s}{\equiv} Y_n$ which in turn implies that $X_n \overset{c}{\equiv} Y_n$. A distribution ensemble $\{X_n\}_{n\in\mathbb{N}}$ is said to be *pseudorandom* if $X_n \overset{c}{\equiv} U_{m(n)}$, where $m(n) = |X_n|$.

2.2.1 Some Useful Facts

We will rely on the following standard properties of statistical distance (proofs for most of these facts can be found in [129]).

Fact 2.1 *For any distributions X, Y, and Z, we have $\|X - Z\| \leq \|X - Y\| + \|Y - Z\|$.*

Fact 2.2 *For any distributions X, X', Y, and Y' we have*

$$\left\|(X \times X') - (Y \times Y')\right\| \leq \|X - Y\| + \|X' - Y'\|,$$

where $A \times B$ denotes the product distribution of A, B, i.e., the joint distribution of independent samples from A and B.

Fact 2.3 *For any distributions X and Y and every (possibly randomized) function A, we have $\|A(X) - A(Y)\| \leq \|X - Y\|$.*

For jointly distributed random variables A and B we write $B|_{A=a}$ to denote the conditional distribution of B given that $A = a$.

Fact 2.4 *Suppose that $X = (X_1, X_2)$ and $Y = (Y_1, Y_2)$ are probability distributions on a set $D \times E$ such that: (1) X_1 and Y_1 are identically distributed; and (2) with probability greater than $1 - \varepsilon$ over $x \leftarrow X_1$, we have $\|X_2|_{X_1=x} - Y_2|_{Y_1=x}\| \le \delta$. Then $\|X - Y\| \le \varepsilon + \delta$.*

Fact 2.5 *Let $\{X_z\}_{z \in \mathcal{Z}}$, $\{Y_z\}_{z \in \mathcal{Z}}$ be distribution ensembles. Then, for every distribution Z over $\mathcal{Z}$, we have $\|(Z, X_Z) - (Z, Y_Z)\| = \mathbb{E}_{z \leftarrow Z}[\|X_z - Y_z\|]$. In particular, if $\|X_z - Y_z\| \le \varepsilon$ for every $z \in \mathcal{Z}$, then $\|(Z, X_Z) - (Z, Y_Z)\| \le \varepsilon$.*

We will also rely on several standard facts about computational indistinguishability (cf. [70, Chap. 2]). We begin with the computational versions of Facts 2.1, 2.2, and 2.3.

Fact 2.6 *For any distribution ensembles X, Y and Z, if $X \overset{c}{\equiv} Y$ and $Y \overset{c}{\equiv} Z$ then $X \overset{c}{\equiv} Z$.*

Fact 2.7 *Let $\{X_n\}, \{X_n'\}, \{Y_n\}$ and $\{Y_n'\}$ be distribution ensembles. Suppose that $X_n \overset{c}{\equiv} Y_n$ and $X_n' \overset{c}{\equiv} Y_n'$. Then $(X_n \times X_n') \overset{c}{\equiv} (Y_n \times Y_n')$, where $A \times B$ denotes the product distribution of A, B (i.e., the joint distribution of independent samples from A and B).*

Fact 2.8 *Suppose that the distribution ensembles $\{X_n\}$ and $\{Y_n\}$ are computationally indistinguishable. Then for every polynomial-time computable function f we have $f(X_n) \overset{c}{\equiv} f(Y_n)$.*

Consider a case in which two probabilistic (possibly computationally unbounded) algorithms behave "similarly" on every input, in the sense that their output distributions are computationally indistinguishable. The following two facts deal with such a situation. Fact 2.9 asserts that an efficient procedure that gets an oracle access to one of these algorithms cannot tell which algorithm it communicates with. Fact 2.10 asserts that the outputs of these algorithms cannot be distinguished with respect to any (not necessarily efficiently samplable) input distribution.

Fact 2.9 *Let X and Y be probabilistic algorithms such that for every string family $\{z_n\}$ where $z_n \in \{0, 1\}^n$, it holds that $X(z_n) \overset{c}{\equiv} Y(z_n)$. Then, for any (non-uniform) polynomial-time oracle machine A, it holds that $A^X(1^n) \overset{c}{\equiv} A^Y(1^n)$ (where A does not have access to the random coins of the given probabilistic oracle).*

Fact 2.10 *Let X and Y be probabilistic algorithms such that for every string family $\{z_n\}$ where $z_n \in \{0, 1\}^n$, it holds that $X(z_n) \overset{c}{\equiv} Y(z_n)$. Then, for every distribution ensemble $\{Z_n\}$ where Z_n is distributed over $\{0, 1\}^n$, we have $(Z_n, X(Z_n)) \overset{c}{\equiv} (Z_n, Y(Z_n))$.*

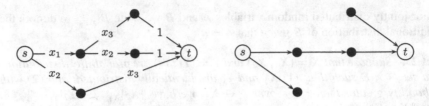

Fig. 2.1 A mod-2 branching program computing the majority of three bits (*left side*), along with the graph G_{110} induced by the assignment 110 (*right side*)

For a randomized algorithm A and an integer i we define A^i to be the randomized algorithm obtained by composing A exactly i times with itself; that is, $A^1(x) = A(x)$ and $A^i(x) = A(A^{i-1}(x))$, where in each invocation a fresh randomness is used. The following fact (which is implicit in [5]) can be proved via a hybrid argument.

Fact 2.11 *Let* $\{X_n\}$ *be a distribution ensemble, and A be a randomized polynomial-time algorithm. Suppose that* $\{X_n\} \stackrel{c}{\equiv} \{A(X_n)\}$. *Then, for every polynomial $p(\cdot)$, we have* $\{X_n\} \stackrel{c}{\equiv} \{A^{p(n)}(X_n)\}$.

2.3 Computational Models

Branching Programs A branching program (BP) is defined by a tuple $BP = (G, \phi, s, t)$, where $G = (V, E)$ is a directed acyclic graph, ϕ is a labeling function assigning each edge either a positive literal x_i, a negative literal $\bar{x}_i$ or the constant 1, and s, t are two distinguished nodes of G. The *size* of BP is the number of nodes in G. Each input assignment $w = (w_1, \ldots, w_n)$ naturally induces an unlabeled subgraph G_w, whose edges include all $e \in E$ such that $\phi(e)$ is satisfied by w (e.g., an edge labeled x_i is satisfied by w if $w_i = 1$). BPs may be assigned different semantics: in a *non-deterministic* BP, an input w is accepted if G_w contains at least one path from s to t; in a (*counting*) *mod-p* BP, the BP computes the number of paths from s to t modulo p. In this work, we will mostly be interested in mod-2 BPs. An example of a mod-2 BP is given in Fig. 2.1.

Circuits Boolean circuits are defined in a standard way. That is, we define a boolean circuit C as a directed acyclic graph with labeled, ordered vertices of the following types: (1) *input* vertices, each labeled with a literal x_i or $\bar{x}_i$ and having fan-in 0; (2) *gate* vertices, labeled with one of the boolean functions AND, OR and having fan-in 2; (3) *output* vertices, labeled "output" and having fan-in 1 and fan-out 0. The edges of the circuit are referred to as *wires*. A wire that outgoes from an input vertex is called an *input wire*, and a wire that enters an output vertex is called an *output wire*. Any input $x \in \{0, 1\}^n$ assigns a unique *value* to each wire in the natural way. The output value of C, denoted $C(x)$, contains the values of the output wires according to the given predefined order. The *size* of a circuit, denoted $|C|$, is the number of wires in C, and its *depth* is the maximum distance from an input to an output (i.e. the length of the longest directed path in the graph).

We say that $C = \{C_n\}$ is an $\mathbf{NC}^i$ circuit family if for every n, the circuit C_n is of size $\mathrm{poly}(n)$ and depth $O(\log^i(n))$. $\mathbf{AC}^i$ circuits are defined similarly, except that gates are allowed to have unbounded fan-in.

$\mathbf{NC}^i$-Reductions   A circuit with an *oracle* access to a function $g : \{0, 1\}^* \to \{0, 1\}^*$ is a circuit that contains, in addition to the bounded fan-in OR, AND gates, special *oracle gates* with unbounded fan-in that compute the function g. We say that $f : \{0, 1\}^* \to \{0, 1\}^*$ is $\mathbf{NC}^i$ *reducible* to g, and write $f \in \mathbf{NC}^i[g]$, if f can be computed by a uniform family of polynomial size, $O(\log^i n)$ depth circuits with oracle gates to g. (Oracle gates are treated the same as AND/OR gates when defining depth.) Note that if $f \in \mathbf{NC}^i[g]$ and $g \in \mathbf{NC}^j$ then $f \in \mathbf{NC}^{i+j}$.

Function Families and Representations   We associate with a function $f : \{0, 1\}^* \to \{0, 1\}^*$ a function family $\{f_n\}_{n \in \mathbb{N}}$, where f_n is the restriction of f to n-bit inputs. We assume all functions to be length regular, namely their output length depends only on their input length. Hence, we may write $f_n : \{0, 1\}^n \to \{0, 1\}^{l(n)}$. We will represent functions f by families of circuits, branching programs, or vectors of polynomials (where each polynomial is represented by a formal sum of monomials). Whenever f is taken from a uniform class, we assume that its representation is uniform as well. That is, the representation of f_n is generated in time $\mathrm{poly}(n)$ and in particular is of polynomial size. We will often abuse notation and write f instead of f_n even when referring to a function on n bits. We will also write $f : \{0, 1\}^n \to \{0, 1\}^{l(n)}$ to denote the family $\{f_n : \{0, 1\}^n \to \{0, 1\}^{l(n)}\}_{n \in \mathbb{N}}$.

Complexity Classes   For brevity, we use the (somewhat nonstandard) convention that all complexity classes are polynomial-time uniform unless otherwise stated. For instance, $\mathbf{NC}^0$ refers to the class of functions admitting uniform $\mathbf{NC}^0$ circuits, whereas *non-uniform* $\mathbf{NC}^0$ refers to the class of functions admitting non-uniform $\mathbf{NC}^0$ circuits. We let $\mathbf{NL}/poly$ (resp., $\oplus\mathbf{L}/poly$) denote the class of boolean functions computed by a polynomial-time uniform family of nondeterministic (resp., modulo-2) BPs. (Recall that in a uniform family of circuits or branching programs computing f, it should be possible to generate the circuit or branching program computing f_n in time $\mathrm{poly}(n)$.) Equivalently, the class $\mathbf{NL}/poly$ (resp., $\oplus\mathbf{L}/poly$) is the class of functions computed by $\mathbf{NL}$ (resp., $\oplus\mathbf{L}$) Turing machines taking a uniform advice. We extend boolean complexity classes, such as $\mathbf{NL}/poly$ and $\oplus\mathbf{L}/poly$, to include non-boolean functions by letting the representation include $l(n)$ branching programs, one for each output. Uniformity requires that the $l(n)$ branching programs be all generated in time $\mathrm{poly}(n)$. Similarly, we denote by $\mathbf{P}$ (resp. $\mathbf{BPP}$) the class of *functions* that can be computed in polynomial time (resp. probabilistic polynomial time). For instance, a function $f : \{0, 1\}^n \to \{0, 1\}^{\ell(n)}$ is in $\mathbf{BPP}$ if there exists a probabilistic polynomial-time machine A such that for every $x \in \{0, 1\}^n$ it

$$NC^0 \subsetneq AC^0 \subsetneq TC^0 \subseteq NC^1 \subseteq L/poly \subseteq NL/poly, \oplus L/poly \subseteq NC^2 \subseteq NC \subseteq P$$

Fig. 2.2 The relations between complexity classes inside **P**. The classes $\oplus L/poly$ and $NL/poly$ contain the class $L/poly$ and are contained in NC^2. In a non-uniform setting the class $\oplus L/poly$ contains the class $NL/poly$ [142]

holds that $\Pr[A(x) \neq f(x)] \leq 2^{-n}$, where the probability is taken over the internal coin tosses of A. Figure 2.2 describes the relations between complexity classes inside **P**.

2.4 Output Locality, Input Locality and Degree

Output Locality Let $f : \{0, 1\}^n \to \{0, 1\}^l$ be a function. We say that the j-th output bit of f depends on the i-th input bit if there exists an assignment such that flipping the i-th input bit changes the value of the j-th output bit of f. The *output locality* of f is the maximal number of input bits on which an output bit of f depends. We say that the function $f : \{0, 1\}^n \to \{0, 1\}^l$ is *c-local* if its output locality is at most c (i.e., each of its output bits depends on at most c input bits), and that $f : \{0, 1\}^* \to \{0, 1\}^*$ is c-local if for every n the restriction of f to n-bit inputs is c-local.

Locality may be also defined as a syntactic property. Let C be a boolean circuit with n inputs and l outputs. Then, the j-th output wire of C depends on the i-th input wire if there exists a directed path from i to j. The *output locality* of a circuit is the maximal number of input wires on which an output wire depends. Recall that the non-uniform class nonuniform-NC^0 includes all functions $f : \{0, 1\}^n \to \{0, 1\}^{l(n)}$ which are computable by a circuit family $\{C_n\}$ of constant depth, polynomial size and bounded fan-in gates. For a constant c, we define the class nonuniform-NC_c^0 as the class of functions which are computable by a circuit family $\{C_n\}$ of constant depth, polynomial size and bounded fan-in gates whose output locality is at most c. Clearly, the output locality of a nonuniform-NC^0 circuit is constant and therefore we can write nonuniform-$NC^0 = \bigcup_{c \in \mathbb{N}}$ nonuniform-NC_c^0. The complexity classes NC^0 and NC_c^0, which are the uniform versions of the above classes, requires the circuit family $\{C_n\}$ to be polynomial-time constructible.

By definition, every function in nonuniform-NC_c^0 is c-local. Also, any function in NC_c^0 is both c-local and polynomial-time computable. The following proposition asserts that the converse also holds. That is, if a function is both c-local and polynomial-time computable then it is possible to efficiently construct an NC_c^0 circuit that computes it.

Proposition 2.1 *Let* $f : \{0, 1\}^n \to \{0, 1\}^{l(n)}$ *be a c-local function which can be computed in polynomial time. Then,* $f \in NC^0$. *That is,* f *can be computed by* *a constant depth circuit family $C = \{C_n\}$ with bounded fan-in, and there exist a polynomial-time circuit constructor A such that $A(1^n) = C_n$.*

Proof We show that given an oracle to a c-local function f, one can efficiently "learn" an $\mathbf{NC}^0$ circuit that computes f. First note that it suffices to show how to learn an $\mathbf{NC}^0$ circuit for a *boolean* c-local function, as in this case we can view f as a sequence of $l(n)$ such functions $f^{(1)}, \ldots, f^{(l(n))}$, one for each output bit of f, and construct an $\mathbf{NC}^0$ circuit for f by concatenating the $\mathbf{NC}^0$ circuits for $f^{(1)}, \ldots, f^{(l(n))}$.

We now show how to learn the set S^* of variables that influence the output bit of a boolean c-local function g. Given such a procedure we can easily compute an $\mathbf{NC}^0$ circuit for g (in constant time) by first recovering the truth table of g (restricted to the bits in S^*) and then converting it into a circuit via a standard transformation.

Algorithm 1 Learning the set of influencing variables of a c-local function g : $\{0, 1\}^n \to \{0, 1\}$

---

1. Go over all subsets $S \subseteq [n]$ of cardinality $\leq c$, where larger sets are processed *before* smaller ones.

   a. Let T_S be the set of all strings $x \in \{0, 1\}^n$ for which $x_{[n] \setminus S} = 0^{n - |S|}$.
   b. For every input bit $i \in [n]$ test whether there exists a witness $x \in T_S$ for which $g(x) \neq g(x_{\oplus i})$. (That is, x is a witness for the fact that the output of g depends on the i-th input bit.)
   c. If all the tests succeed, output S and terminate.

Let S be the output of the algorithm. We prove that $S = S^*$. First note that $S \subseteq S^*$, as g depends on all the input bits of S (for each $i \in S$ we found a witness for this dependency). Hence, since Algorithm 1 processes larger sets before smaller ones, it does not terminate before it reaches S^*. It is left to show that the algorithm stops when it examines S^*. Fix some $i \in S^*$ and let $x \in \{0, 1\}^n$ be the witness for the fact that g depends on i, namely, $g(x) \neq g(x_{\oplus i})$. Let $y \in \{0, 1\}^n$ be a string for which $y_{S^*} = x_{S^*}$ and $y_{[n] \setminus S^*} = 0^{n - |S^*|}$. Clearly $y \in T_{S^*}$. Also, since the indices outside of S^* do not affect g, we have $g(y) = g(x') \neq g(x_{\oplus i}) = g(y_{\oplus i})$ and the proposition follows. □

Input Locality The *input locality* of a function f is the maximal number of output bits on which an input bit of f has influence. We envision circuits as having their inputs at the bottom and their outputs at the top. Accordingly, for functions $\ell(n), m(n)$, the class nonuniform-$\mathbf{Local}_{\ell(n)}^{m(n)}$ (resp. nonuniform-$\mathbf{Local}_{\ell(n)}$, nonuniform-$\mathbf{Local}^{m(n)}$) includes all functions $f : \{0, 1\}^* \to \{0, 1\}^*$ whose input locality is $\ell(n)$ and whose output locality is $m(n)$ (resp. whose input locality is $\ell(n)$, whose output locality is $m(n)$). The uniform versions of these classes contain only functions that can be computed in polynomial time. (Note that $\mathbf{Local}^{O(1)}$ is equivalent to the class $\mathbf{NC}^0$. However, in most cases we will prefer the notation $\mathbf{NC}^0$.) We now prove the following proposition.

Proposition 2.2 *A function* $f : \{0, 1\}^n \to \{0, 1\}^{l(n)}$ *is in the class* $\mathbf{Local}_{O(1)}^{O(1)}$ *if and only if* f *can be computed by polynomial-time constructible circuit family* $\{C_n\}$ *of constant depth, whose gates have bounded fan-in and bounded fan-out.*

Proof Suppose that $f \in \mathbf{Local}_\ell^m$ for some constants $\ell, m \in \mathbb{N}$. Then, as shown in the proof of Proposition 2.1, we can construct, in time poly(n), an $\mathbf{NC}^0$ circuit C_n that computes f_n. Moreover, the circuit computing each output bit involves only those input bits on which this output depend. Hence, the gates of C_n have also bounded fan-out.

To prove the converse direction, note that if f_n is computable by a circuit C_n with depth d, fan-in b, and fan-out c, where $b, c, d = O(1)$, then the input locality of f is at most $c^d = O(1)$ and its output locality is at most $b^d = O(1)$. Also, if C_n is constructible in time poly(n), then $f \in \mathbf{P}$. $\square$

Most of this book deals with output locality, hence, by default, the term *locality* always refers to *output* locality.

Degree   We will sometimes view the binary alphabet as the finite field $\mathbb{F} = \mathbb{F}_2$, and say that a function $f : \mathbb{F}^n \to \mathbb{F}^{l(n)}$ has degree d if each of its outputs can be expressed as a multivariate polynomial of degree (at most) d in the inputs. The output locality of a function trivially upper bounds its degree.

Chapter 3
Randomized Encoding of Functions

Abstract In this chapter we formally introduce our notion of randomized encoding which will be used as a central tool in subsequent chapters. In Sect. 3.1 we introduce several variants of randomized encoding and in Sect. 3.2 we prove some of their useful properties. Finally, in Sect. 3.3 we discuss other aspects and limitations of randomized encoding such as the necessity of randomness, the expressive power of encoding in $\mathbf{NC}^0$, and the lowest output locality which is sufficient to encode all functions.

3.1 Definitions

We start by defining a randomized encoding of a finite function f. This definition will be later extended to a (uniform) family of functions.

Definition 3.1 (Randomized encoding) Let $f : \{0, 1\}^n \to \{0, 1\}^l$ be a function. We say that a function $\hat{f} : \{0, 1\}^n \times \{0, 1\}^m \to \{0, 1\}^s$ is a δ-correct, ε-private *randomized encoding* of f, if it satisfies the following:

- δ-**correctness**. There exists a deterministic[1] algorithm B, called a *decoder*, such that for every input $x \in \{0, 1\}^n$, $\Pr[B(\hat{f}(x, U_m)) \neq f(x)] \leq \delta$.
- ε-**privacy**. There exists a randomized algorithm S, called a *simulator*, such that for every $x \in \{0, 1\}^n$, $\|S(f(x)) - \hat{f}(x, U_m)\| \leq \varepsilon$.

We refer to the second input of $\hat{f}$ as its *random input* and to m and s as the *randomness complexity* and *output complexity* of $\hat{f}$, respectively.

Note that the above definition only refers to the *information* about x revealed by $\hat{f}(x, r)$ and does not consider the complexity of the decoder and the simulator. Intuitively, the function $\hat{f}$ defines an "information-theoretically equivalent" representation of f. The correctness property guarantees that from $\hat{y} = \hat{f}(x, r)$ it is possible

---

[1] We restrict the decoder to be deterministic for simplicity. This restriction does not compromise generality, in the sense that one can transform a randomized decoder to a deterministic one by incorporating the coins of the former in the encoding itself.

B. Applebaum, *Cryptography in Constant Parallel Time*,
Information Security and Cryptography,
DOI 10.1007/978-3-642-17367-7_3, © Springer-Verlag Berlin Heidelberg 2014

to reconstruct $f(x)$ (with high probability), whereas the privacy property guarantees that by seeing $\hat{y}$ one cannot learn too much about x (in addition to $f(x)$). The encoding is δ-correct (resp. ε-private), if it is correct (resp. private) up to an "error" of δ (resp., ε).

In general, we will be interested in encodings $\hat{f}$ which are "cheaper" then the original function f under some complexity measure such as output locality, input locality, or algebraic degree. This is illustrated by the next example.

Example 3.1 Consider the OR_n function $f(x_1, \ldots, x_n) = x_1 \vee x_2 \vee \cdots \vee x_n$. We define a randomized encoding $\hat{f} : \{0, 1\}^n \times \{0, 1\}^{ns} \to \{0, 1\}^s$ by

$$\hat{f}(x, r) = \left(\sum_{i=1}^{n} x_i r_{i,1}, \ldots, \sum_{i=1}^{n} x_i r_{i,s} \right),$$

where $x = (x_1, \ldots, x_n)$, $r = (r_{i,j})$ for $1 \leq i \leq n$, $1 \leq j \leq s$, and addition and multiplication are over $\mathbb{F}_2$. First, observe that the distribution of $\hat{f}(x, U_{ns})$ depends only on the value of $f(x)$. Specifically, let S be a simulator that outputs an s-tuple of zeroes if $f(x) = 0$, and a uniformly chosen string in $\{0, 1\}^s$ if $f(x) = 1$. It is easy to verify that $S(f(x))$ is distributed the same as $\hat{f}(x, U_{ns})$ for any $x \in \{0, 1\}^n$. It follows that this randomized encoding is 0-private. Also, one can obtain an efficient decoder B that given a sample y from the distribution $\hat{f}(x, U_{ns})$ outputs 0 if $y = 0^s$ and otherwise outputs 1. Such an algorithm will err with probability 2^{-s}, thus $\hat{f}$ is 2^{-s}-correct. Finally, observe that the algebraic degree of f is n, whereas $\hat{f}$ is a degree 2 encoding.

On Uniform Randomized Encodings The above definition naturally extends to functions $f : \{0, 1\}^* \to \{0, 1\}^*$. In this case, the parameters $l, m, s, \delta, \varepsilon$ are all viewed as functions of the input length n, and the algorithms B, S receive 1^n as an additional input. In our default uniform setting, we require that $\hat{f}_n$, the encoding of f_n, be computable in time poly(n) (given $x \in \{0, 1\}^n$ and $r \in \{0, 1\}^{m(n)}$). Thus, in this setting both $m(n)$ and $s(n)$ are polynomially bounded. We also require both the decoder and the simulator to be efficient. (This is not needed by some of the applications, but is a feature of our constructions.) We formalize these requirements below.

Definition 3.2 (Uniform randomized encoding) Let $f : \{0, 1\}^* \to \{0, 1\}^*$ be a polynomial-time computable function and $l(n)$ an output length function such that $|f(x)| = l(|x|)$ for every $x \in \{0, 1\}^*$. We say that $\hat{f} : \{0, 1\}^* \times \{0, 1\}^* \to \{0, 1\}^*$ is a $\delta(n)$-correct $\epsilon(n)$-private uniform randomized encoding of f, if the following holds:

- **Length regularity**. There exist polynomially-bounded and efficiently computable length functions $m(n), s(n)$ such that for every $x \in \{0, 1\}^n$ and $r \in \{0, 1\}^{m(n)}$, we have $|\hat{f}(x, r)| = s(n)$.
- **Efficient evaluation**. There exists a polynomial-time *evaluation algorithm* that, given $x \in \{0, 1\}^*$ and $r \in \{0, 1\}^{m(|x|)}$, outputs $\hat{f}(x, r)$.

- δ-**correctness**. There exists a polynomial-time *decoder* B, such that for every $x \in \{0, 1\}^n$ we have $\Pr[B(1^n, \hat{f}(x, U_{m(n)})) \neq f(x)] \leq \delta(n)$.
- ε-**privacy**. There exists a probabilistic polynomial-time *simulator* S, such that for every $x \in \{0, 1\}^n$ we have $\|S(1^n, f(x)) - \hat{f}(x, U_{m(n)})\| \leq \varepsilon(n)$.

When saying that a uniform encoding $\hat{f}$ is in a (uniform) circuit complexity class, we mean that its evaluation algorithm can be implemented by circuits in this class. For instance, we say that $\hat{f}$ is in $\mathbf{NC}_d^0$ if there exists a polynomial-time circuit generator G such that $G(1^n)$ outputs a d-local circuit computing $\hat{f}(x, r)$ on all $x \in \{0, 1\}^n$ and $r \in \{0, 1\}^{m(n)}$.

From here on, a randomized encoding of an efficiently computable function is assumed to be uniform by default. Moreover, we will freely extend the above definition to apply to a uniform collection of functions $\mathscr{F} = \{f_z\}_{z \in Z}$, for some index set $Z \subseteq \{0, 1\}^*$. In such a case it is required that the encoded collection $\hat{\mathscr{F}} = \{\hat{f}_z\}_{z \in Z}$ is also uniform, in the sense that the same efficient evaluation algorithm, decoder, and simulator should apply to the entire collection when given z as an additional input. (See Appendix 4.9 for a more detailed discussion of *collections* of functions and cryptographic primitives.) Finally, for the sake of simplicity we will sometimes formulate our definitions, claims and proofs using finite functions, under the implicit understanding that they naturally extend to the uniform setting.

We move on to discuss some variants of the basic definition. Correctness (resp., privacy) can be either *perfect*, when $\delta = 0$ (resp., $\varepsilon = 0$), or *statistical*, when $\delta(n)$ (resp., $\varepsilon(n)$) is negligible. In fact, we can further relax privacy to hold only against efficient adversaries, e.g., to require that for every $x \in \{0, 1\}^n$, every polynomial-size circuit family $\{A_n\}$ distinguishes between the distributions $S(f(x))$ and $\hat{f}(x, U_m)$ with no more than negligible advantage. Such an encoding is referred to as *computationally* private and it suffices for the purpose of many applications. However, while for some of the primitives (such as OWF) computational privacy and statistical correctness will do, others (such as PRGs or one-way permutations) require even stronger properties than perfect correctness and privacy. One such additional property is that the simulator S, when invoked on a uniformly random string from $\{0, 1\}^l$ (the output domain of f), will output a uniformly random string from $\{0, 1\}^s$ (the output domain of $\hat{f}$). We call this property *balance*. Note that the balance requirement does not impose any uniformity condition on the output of f, which in fact can be concentrated on a strict subset of $\{0, 1\}^l$.

Definition 3.3 (Balanced randomized encoding) A randomized encoding $\hat{f} : \{0, 1\}^n \times \{0, 1\}^m \to \{0, 1\}^s$ of a function $f : \{0, 1\}^n \to \{0, 1\}^l$ is called *balanced* if it has a perfectly private simulator S such that $S(U_l) \equiv U_s$. We refer to S as a *balanced simulator*.

A last useful property is a syntactic one: we sometimes want $\hat{f}$ to have the same additive stretch as f. Specifically, we say that $\hat{f}$ is *stretch-preserving* (with respect to f) if $s - (n + m) = l - n$, or equivalently $m = s - l$.

We are now ready to define our three main variants of randomized encoding.

Definition 3.4 (Perfect randomized encoding) A *perfect randomized encoding* is a randomized encoding that is perfectly correct, perfectly private, balanced, and stretch-preserving.

Definition 3.5 (Statistical randomized encoding) A *statistical randomized encoding* is a randomized encoding that is statistically correct and statistically private.

Definition 3.6 (Computational randomized encoding) Let $f = \{f_n : \{0, 1\}^n \to \{0, 1\}^{\ell(n)}\}$ be a function family. We say that the function family $\hat{f} = \{\hat{f}_n : \{0, 1\}^n \times \{0, 1\}^{m(n)} \to \{0, 1\}^{s(n)}\}$ is a *computational randomized encoding* of f (or computational encoding for short), if it satisfies the following requirements:

- **Statistical correctness**. There exists a polynomial-time decoder B, such that for every $x \in \{0, 1\}^n$, we have $\Pr[B(1^n, \hat{f}_n(x, U_{m(n)})) \neq f_n(x)] \leq \delta(n)$, for some negligible function $\delta(n)$.
- **Computational privacy**. There exists a probabilistic polynomial-time simulator S, such that for any family of strings $\{x_n\}_{n \in \mathbb{N}}$ where $|x_n| = n$, we have $S(1^n, f_n(x_n)) \overset{c}{\equiv} \hat{f}_n(x_n, U_{m(n)})$.

We will also refer to *perfectly correct* computational encodings, where the statistical correctness requirement is strengthened to perfect correctness. (In fact, the construction of Sect. 5.2 yields such an encoding.)

A Combinatorial View of Perfect Encoding To gain better understanding of the properties of perfect encoding, we take a closer look at the relation between a function and its encoding. Let $\hat{f} : \{0, 1\}^{n+m} \to \{0, 1\}^s$ be an encoding of $f : \{0, 1\}^n \to \{0, 1\}^l$. The following description addresses the simpler case where f is onto. Every $x \in \{0, 1\}^n$ is mapped to some $y \in \{0, 1\}^l$ by f, and to a 2^m-size multiset $\{\hat{f}(x, r) | r \in \{0, 1\}^m\}$ which is contained in $\{0, 1\}^s$. Perfect privacy means that this multiset is common to all the x's that share the same image under f; so we have a mapping from $y \in \{0, 1\}^l$ to multisets in $\{0, 1\}^s$ of size 2^m (such a mapping is defined by the perfect simulator). Perfect correctness means that these multisets are mutually disjoint. However, even perfect privacy and perfect correctness together do not promise that this mapping covers all of $\{0, 1\}^s$. The balance property guarantees that the multisets form a perfect tiling of $\{0, 1\}^s$; moreover it promises that each element in these multisets has the same multiplicity. If the encoding is also stretch-preserving, then the multiplicity of each element must be 1, so that the multisets are actually sets. Hence, a perfect randomized encoding guarantees the existence of a perfect simulator S whose 2^l output distributions form a perfect tiling of the space $\{0, 1\}^s$ by sets of size 2^m.

Remark 3.1 (**A padding convention**) We will sometimes view $\hat{f}$ as a function of a single input of length $n + m(n)$ (e.g., when using it as an OWF or a PRG). In this case, we require $m(\cdot)$ to be monotone non-decreasing, so that $n + m(n)$ uniquely determines n. We apply a standard padding technique for defining $\hat{f}$ on inputs whose length is not of the form $n + m(n)$. Specifically, if $n + m(n) + t < (n+1) + m(n+1)$ we define $\hat{f}'$ on inputs of length $n + m(n) + t$ by applying $\hat{f}_n$ on the first $n + m(n)$ bits and then appending the t additional input bits to the output of $\hat{f}_n$. This convention respects the security of cryptographic primitives such as OWF, PRG, and collision-resistant hashing, provided that $m(n)$ is efficiently computable and is sufficiently dense (both of which are guaranteed by a uniform encoding). That is, if the unpadded function $\hat{f}$ is secure with respect to its partial domain, then its padded version $\hat{f}'$ is secure in the standard sense, i.e., over the domain of all strings.[2] (See a proof for the case of OWF in [70, Proposition 2.2.3].) Note that the padded function $\hat{f}'$ has the same locality and degree as $\hat{f}$. Moreover, $\hat{f}'$ also preserves syntactic properties of $\hat{f}$; for example it preserves the stretch of $\hat{f}$, and if $\hat{f}$ is a permutation then so is $\hat{f}'$. Thus, it is enough to prove our results for the partially defined unpadded function $\hat{f}$, and keep the above conventions implicit.

Finally, we define three complexity classes that capture the power of randomized encodings in NC^0.

Definition 3.7 (The classes PREN, SREN, CREN) The class **PREN** (resp., **SREN, CREN**) is the class of functions $f : \{0,1\}^* \to \{0,1\}^*$ admitting a perfect (resp., statistical, computational) uniform randomized encoding in NC^0. (As usual, NC^0 is polynomial-time uniform.)

It follows from the definitions that **PREN** $\subseteq$ **SREN** $\subseteq$ **CREN**.[3] We will later show that $\oplus L/poly \subseteq$ **PREN** (Theorem 4.1), **NL**/$poly \subseteq$ **SREN** (Theorem 4.2), and, assuming that there exists a OWF in **SREN**, the class **CREN** is equal to the class **BPP** (Theorem 5.2). Moreover, functions in all of these classes admit an encoding of degree 3 and (output) locality 4 (Corollary 4.1).

3.2 Basic Properties

We now put forward some useful properties of randomized encodings. We first argue that an encoding of a non-boolean function can be obtained by concatenating encodings of its output bits, using an independent random input for each bit. The resulting encoding inherits all the features of the concatenated encodings, and in particular preserves their perfectness.

---

[2]This can be generally explained by viewing each slice of the padded function $\hat{f}'$ (i.e., its restriction to inputs of some fixed length) as a *perfect* randomized encoding of a corresponding slice of $\hat{f}$.

[3]We also note that if **CREN** $\neq$ **SREN** then there exist two polynomial-time constructible ensembles which are computationally indistinguishable but not statistically close. In [67] it is shown that such ensembles implies the existence of infinitely often OWF, i.e., a polynomial-time computable function which is hard to invert for infinitely many input lengths (see [70, Definition 4.5.4] for a formal definition).

Lemma 3.1 (Concatenation) *Let $f_i : \{0,1\}^n \rightarrow \{0,1\}$, $1 \leq i \leq l$, be the boolean functions computing the output bits of a function $f : \{0,1\}^n \rightarrow \{0,1\}^l$. If $\hat{f}_i : \{0,1\}^n \times \{0,1\}^{m_i} \rightarrow \{0,1\}^{s_i}$ is a δ-correct ε-private encoding of f_i, then the function $\hat{f} : \{0,1\}^n \times \{0,1\}^{m_1 + \cdots + m_l} \rightarrow \{0,1\}^{s_1 + \cdots + s_l}$ defined by $\hat{f}(x, (r_1, \ldots, r_l)) \stackrel{\text{def}}{=} (\hat{f}_1(x, r_1), \ldots, \hat{f}_l(x, r_l))$ is a (δl)-correct, (εl)-private encoding of f. Moreover, if all $\hat{f}_i$ are perfect then so is $\hat{f}$.*

Proof We start with correctness. Let B_i be a δ-correct decoder for $\hat{f}_i$. Define a decoder B for $\hat{f}$ by $B(\hat{y}_1, \ldots, \hat{y}_l) = (B_1(\hat{y}_1), \ldots, B_l(\hat{y}_l))$. By a union bound argument, B is a (δl)-correct decoder for $\hat{f}$ as required.

We turn to analyze privacy. Let S_i be an ε-private simulator for $\hat{f}_i$. An (εl)-private simulator S for $\hat{f}$ can be naturally defined by $S(y) = (S_1(y_1), \ldots, S_l(y_l))$, where the invocations of the simulators S_i use independent coins. Indeed, for every $x \in \{0,1\}^n$ we have:

$$\left\| S(f(x)) - \hat{f}\big(x, (U_{m_1}, \ldots, U_{m_l})\big) \right\| = \left\| \big(S_i(y_i)\big)_{i=1}^l - \big(\hat{f}_i(x, U_{m_i})\big)_{i=1}^l \right\|$$

$$\leq \sum_{i=1}^l \left\| S_i(y_i) - \hat{f}_i(x, U_{m_i}) \right\|$$

$$\leq \varepsilon l,$$

where $y = f(x)$. The first inequality follows from Fact 2.2 and the independence of the randomness used for different i, and the second from the ε-privacy of each S_i.

Note that the simulator S described above is balanced if all S_i are balanced. Moreover, if all $\hat{f}_i$ are stretch preserving, i.e., $s_i - 1 = m_i$, then we have $\sum_{i=1}^l s_i - l = \sum_{i=1}^l m_i$ and hence $\hat{f}$ is also stretch preserving. It follows that if all $\hat{f}_i$ are perfect then so is $\hat{f}$.   □

We state the following uniform version of Lemma 3.1, whose proof is implicit in the above.

Lemma 3.2 (Concatenation: uniform version) *Let $f : \{0,1\}^* \rightarrow \{0,1\}^*$ be a polynomial-time computable function, viewed as a uniform collection of functions $\mathscr{F} = \{f_{n,i}\}_{n \in \mathbb{N}, 1 \leq i \leq l(n)}$; that is, $f_{n,i}(x)$ outputs the i-th bit of $f(x)$ for all $x \in \{0,1\}^n$. Suppose that $\hat{\mathscr{F}} = \{\hat{f}_{n,i}\}_{n \in \mathbb{N}, 1 \leq i \leq l(n)}$ is a perfect (resp., statistical) uniform randomized encoding of $\mathscr{F}$. Then, the function $\hat{f} : \{0,1\}^* \times \{0,1\}^* \rightarrow \{0,1\}^*$ defined by $\hat{f}(x, (r_1, \ldots, r_{l(|x|)})) \stackrel{\text{def}}{=} (\hat{f}_{|x|,1}(x, r_1), \ldots, \hat{f}_{|x|, l(|x|)}(x, r_{l(|x|)}))$ is a perfect (resp., statistical) uniform randomized encoding of f.*

The concatenation lemma can be extended to the computational setting. However, since this version is not going to be used anywhere in this book and since the formal statement is somewhat technical, we chose to omit it.

Another useful feature of randomized encodings is the following intuitive composition property: suppose we encode f by g, and then view g as a deterministic function and encode it again. Then, the resulting function (parsed appropriately) is a randomized encoding of f. Again, the resulting encoding inherits the perfectness of the encodings from which it is composed.

Lemma 3.3 (Composition) *Let $g(x, r_g)$ be a δ_g-correct, ε_g-private encoding of $f(x)$ and $h((x, r_g), r_h)$ be a δ_h-correct, ε_h-private encoding of $g((x, r_g))$ (viewed as a single-argument function). Then, the function $\hat{f}(x, (r_g, r_h)) \stackrel{\text{def}}{=} h((x, r_g), r_h)$ is a $(\delta_g + \delta_h)$-correct, $(\varepsilon_g + \varepsilon_h)$-private encoding of f. Moreover, if g, h are perfect (resp., statistical) uniform randomized encodings then so is $\hat{f}$.*

Proof We start with correctness. Let B_g be a δ_g-correct decoder for g and B_h a δ_h-correct decoder for h. Define a decoder B for $\hat{f}$ by $B(\hat{y}) = B_g(B_h(\hat{y}))$. The decoder B errs only if either B_h or B_g err. Thus, by the union bound we have for every x,

$$\Pr_{r_g, r_h}\left[B\left(\hat{f}\left(x, (r_g, r_h)\right)\right) \neq f(x)\right] \leq \Pr_{r_g, r_h}\left[B_h\left(h((x, r_g), r_h)\right) \neq g(x, r_g)\right]$$
$$+ \Pr_{r_g}\left[B_g\left(g(x, r_g)\right) \neq f(x)\right]$$
$$\leq \delta_h + \delta_g,$$

as required.

Privacy is argued similarly. Let S_g be an ε_g-private simulator for g and S_h an ε_h-private simulator for h. We define a simulator S for $\hat{f}$ by $S(y) = S_h(S_g(y))$. Letting m_g, m_h denote the randomness complexity of g, h, respectively, we have for every x,

$$\left\| S(f(x)) - \hat{f}\left(x, (U_{m_g}, U_{m_h})\right) \right\| = \left\| S_h\left(S_g(f(x))\right) - h\left((x, U_{m_g}), U_{m_h}\right) \right\|$$
$$\leq \left\| S_h\left(S_g(f(x))\right) - S_h\left(g(x, U_{m_g})\right) \right\|$$
$$+ \left\| S_h\left(g(x, U_{m_h})\right) - h\left((x, U_{m_g}), U_{m_h}\right) \right\|$$
$$\leq \varepsilon_g + \varepsilon_h,$$

where the first inequality follows from the triangle inequality (Fact 2.1), and the second from Facts 2.3 and 2.5.

It is easy to verify that if S_g and S_h are balanced then so is S. Moreover, if g preserves the additive stretch of f and h preserves the additive stretch of g then h (hence also $\hat{f}$) preserves the additive stretch of f. Thus $\hat{f}$ is perfect if both g, h are perfect. All the above naturally carries over to the uniform setting, from which the last part of the lemma follows. □

The composition lemma can be easily extended to the computational setting.

Lemma 3.4 (Composition of computational encoding) *Let $g(x, r_g)$ be a computational encoding of $f(x)$ and $h((x, r_g), r_h)$ a computational encoding of $g((x, r_g))$,* *viewing the latter as a single-argument function. Then, the function $\hat{f}(x, (r_g, r_h)) \overset{\text{def}}{=} h((x, r_g), r_h)$ is a computational encoding of $f(x)$ whose random inputs are (r_g, r_h). Moreover, if g, h are perfectly correct then so is $\hat{f}$.*

Proof Correctness follows from the same arguments as in the proof of Lemma 3.3. To prove computational privacy, we again define a simulator S for $\hat{f}$ by $S(y) = S_h(S_g(y))$, where S_g (resp., S_h) is a computationally-private simulator for g (resp., h). Letting $m_g(n)$ and $m_h(n)$ denote the randomness complexity of g and h, respectively, and $\{x_n\}_{n \in \mathbb{N}}$ be a family of strings where $|x_n| = n$, we have,

$$S_h\big(S_g(f(x_n))\big) \overset{c}{\equiv} S_h\big(g(x_n, U_{m_g(n)})\big) \qquad \text{(comp. privacy of } g, \text{ Fact 2.8)}$$

$$\overset{c}{\equiv} h\big((x_n, U_{m_g(n)}), U_{m_h(n)}\big) \quad \text{(comp. privacy of } h, \text{ Fact 2.10).}$$

Hence, the transitivity of the relation $\overset{c}{\equiv}$ (Fact 2.6) completes the proof. $\qquad\square$

It follows as a special case that the composition of a computational encoding with a perfect or a statistical encoding is a computational encoding.

Finally, we prove two useful features of a *perfect* encoding.

Lemma 3.5 (Unique randomness) *Suppose $\hat{f}$ is a perfect randomized encoding of f. Then, (a) $\hat{f}$ satisfies the following* unique randomness *property: for any input x, the function $\hat{f}(x, \cdot)$ is injective, namely there are no distinct r, r' such that $\hat{f}(x, r) = \hat{f}(x, r')$. Moreover, (b) if f is a permutation then so is $\hat{f}$.*

Proof Let $f : \{0, 1\}^n \to \{0, 1\}^l$ and $\hat{f} : \{0, 1\}^n \times \{0, 1\}^m \to \{0, 1\}^s$. To prove part (a), assume towards a contradiction that $\hat{f}$ does not satisfy the unique randomness property. Then, by perfect privacy, we have $|\text{Im}(\hat{f})| < |\text{Im}(f)| \cdot 2^m$. On the other hand, letting S be a balanced simulator, we have

$$|\text{Im}(\hat{f})| \cdot 2^{-s} = \Pr_{y \leftarrow U_l}\big[S(y) \in \text{Im}(\hat{f})\big]$$

$$\geq \Pr_{y \leftarrow U_l}\big[S(y) \in \text{Im}(\hat{f})\big| y \in \text{Im}(f)\big] \cdot \Pr_{y \leftarrow U_l}\big[y \in \text{Im}(f)\big]$$

$$= 1 \cdot \frac{|\text{Im}(f)|}{2^l},$$

where the last equality follows from perfect privacy. Since g is stretch preserving $(s - l = m)$, we get from the above that $|\text{Im}(\hat{f})| \geq |\text{Im}(f)| \cdot 2^m$, and derive a contradiction.

If f is a permutation then $n = l$ and since $\hat{f}$ is stretch preserving, we can write $\hat{f} : \{0, 1\}^s \to \{0, 1\}^s$. Thus, to prove part (b), it is enough to prove that $\hat{f}$ is injective. Suppose that $\hat{f}(x, r) = \hat{f}(x', r')$. Then, since f is injective and $\hat{f}$ is perfectly correct it follows that $x = x'$; hence, by part (a), $r = r'$ and the proof follows. $\qquad\square$

3.3 More Aspects of Randomized Encoding

In the following we address some natural questions regarding the complexity of randomized encoding. The reader is referred to [9] for a survey on the applications of this notion (beyond parallel-time cryptography).

3.3.1 The Necessity of Randomness

We begin by asking whether randomization is really needed in order to encode a function f by a function $\hat{f}$ with constant output locality (i.e., $\mathbf{NC}^0$ function) or constant input locality. The following observation shows that, at least when f is a boolean function, only trivial functions admit such an encoding.

Observation 3.1 *If* $f : \{0, 1\}^n \to \{0, 1\}$ *can be encoded by a function* $\hat{f} : \{0, 1\}^n \to \{0, 1\}^s$ *whose output locality* (*resp. input locality*) *is* c, *then the output locality* (*resp. input locality*) *of* f *is* c.

Proof The privacy of the encoding promises that there exists a pair of strings $y_0, y_1 \in \{0, 1\}^s$ such that for every $x \in \{0, 1\}^n$ we have $\hat{f}(x) = y_{f(x)}$. Also, by perfect correctness, $y_0 \neq y_1$. Assume, without loss of generality, that y_0 and y_1 differ in the first bit. Then, we can compute $f(x)$ by computing the first bit of $\hat{f}(x)$ or its negation. Thus, the output locality and input locality of f are equal to those of $\hat{f}$. $\square$

The above argument relies heavily on the fact that f is a boolean function. Indeed, the claim does not hold in the case of non-boolean functions. Suppose, for example, that $f : \{0, 1\}^n \to \{0, 1\}^n$ is a permutation. Then it can be trivially encoded in $\mathbf{Local}_1^1$ by the identity function. Moreover, if f can be computed and inverted in polynomial time, then the encoding allows efficient decoding and simulation.

3.3.2 The Power of Randomized Encoding

3.3.2.1 Statistical Encoding for Functions Outside of NC

It is very interesting to find out what is the power of randomized encoding in $\mathbf{NC}^0$; that is, which functions are in $\mathbf{SREN}$. Since our general constructions of statistical randomized encoding apply only to various logspace classes (e.g., $\mathbf{NL}/poly$ or $\oplus\mathbf{L}/poly$), one might suspect that $\mathbf{SREN}$ is limited to such functions. It turns out that there are functions in $\mathbf{SREN}$ that are not known to be computable even in $\mathbf{NC}$. Consider, for example, the collection of boolean functions $\{QR_p(x)\}$, which check whether x is a quadratic residue modulo a prime p (i.e., this collection is indexed by primes). This collection can be encoded by the collection $\hat{QR}_p(x, r) = xr^2 \bmod p$

where r is uniformly chosen from $[p]$. Since $\{\hat{QR}_p\}$ can be approximated (up to a negligible error) by a family of $\mathbf{NC}^1$ circuits (over a binary alphabet), we can statistically encode it by an $\mathbf{NC}^0$ collection, which, by composition (Lemma 3.3), also encodes the collection $\{QR_p\}$. However, the collection $\{QR_p\}$ is not known to be computable in $\mathbf{NC}$.

3.3.2.2 Encoding with Unbounded Decoder and Simulator

For some applications, we can relax the uniformity of randomized encoding and allow the decoder and/or the simulator to be computationally unbounded.[4] For example, a perfect encoding with an unbounded simulator and an unbounded decoder still preserves the security of collision-resistant hashing. Moreover, the security of several primitives is preserved by an encoding with an efficient simulator and an unbounded decoder (see Table 4.2). Such variants are also useful for information theoretic secure computation between computationally unbounded parties (cf. [25]). Hence, we would like to understand the power of randomized encodings in this setting too.

Of course, whenever the decoder of the encoding is restricted to run in polynomial time we cannot hope to represent functions that are not efficiently computable (i.e., out of **BPP**). However, it turns out that if we do not restrict the running time of the decoder, some functions that are assumed to be intractable can be encoded by $\mathbf{NC}^0$ functions. For example, consider the function GI that given a graph G (represented as an $n \times n$ adjacency matrix), outputs the lexicographically first graph H that is isomorphic to G. This function is not known to be computable by a polynomial-time algorithm (as such an algorithm would imply that the graph isomorphism language is in **P**). However, we can encode GI by the function $\hat{GI}(G, r) = M(r)GM^T(r)$ where $M(\cdot)$ is a mapping from random bits to an (almost) uniformly chosen $n \times n$ permutation matrix. It is not hard to verify that $\hat{GI}$ is a statistical encoding of GI. Also, the encoding $\hat{GI}$ can be computed in $\mathbf{NC}^1$, and therefore GI can be encoded by an $\mathbf{NC}^0$ encoding (with an efficient simulator). A similar example (for a function that is not known to be efficiently computable but can be represented by an $\mathbf{NC}^0$ encoding) can be obtained by a variant of the quadratic residuosity function presented above in which the modulus is a composite.

On the other hand, the following observation shows that "hard" functions are unlikely to have efficiently computable encodings, even if we allow non-efficient simulation and decoding.

Observation 3.2 *Let f be a boolean function, L be the language that corresponds to f (i.e., $L = f^{-1}(1)$) and $\hat{f}$ be a polynomial-time computable encoding of f. Then,*

---

[4]The encoding itself should still be computable in (polynomial-time) uniform $\mathbf{NC}^0$.

1. *If the encoding is perfect and the simulator is efficient then $L \in \mathbf{NP} \cap \mathbf{coNP}$.*
2. *If the encoding is perfect (and the simulator is not necessarily efficient) then $L \in$ (non-uniform) $\mathbf{NP}/poly \cap$ (non-uniform) $\mathbf{coNP}/poly$.*
3. *If the encoding is statistical and the simulator is efficient then L has a statistical zero-knowledge proof system.*
4. *If the encoding is statistical (and the simulator is not necessarily efficient) then L has a statistical zero-knowledge proof system with a non-uniform verifier.*

Proof Sketch 1. We use the randomness of the encoding as a witness; namely, $f(x) = b$ iff $\exists r$ such that $S_{\bar{0}}(b) = \hat{f}(x, r)$ (i.e., we fix the random coins of the simulator to be the all-zero string.)

2. We use the same argument as in (1), but instead of computing $S_{\bar{0}}(b)$ we use the advice $\hat{f}(x_b, \bar{0})$, where $\bar{0}$ is the all-zero string and x_b is a fixed n-bit string that satisfies $f(x_b) = b$.

3. The protocol is very similar to the standard zero-knowledge protocol of graph non-isomorphism [81]; namely, the verifier selects randomly $b \in \{0, 1\}$: if $b = 1$ it sends a sample from $\hat{f}(x, r)$, otherwise it sends a sample from $S(0)$. The prover has to find the bit b; if the prover succeeds then the verifier accepts, otherwise the verifier rejects. To see that this is indeed a ZK-protocol (against an honest-verifier), note that $f(x) = 1$ implies that the distributions $\hat{f}(x, r)$ and $S(0)$ are almost disjoint, and $f(x) = 0$ implies that these distributions are statistically close.[5]

4. We use the same argument as in (3), but instead of computing $S(0)$ directly, we will sample $\hat{f}(x_0, r)$, where $x_0 \in \{0, 1\}^n$ is an advice which satisfies $f(x_0) = 0$. By statistical privacy, the latter distribution approximates $S(0)$ up to a negligible statistical error. $\square$

As a corollary we can deduce that in all the above settings one is unlikely to obtain a polynomial-time computable encoding for an **NP**-hard language [2, 40, 63, 146].

3.3.3 Lower Bounds for Locality and Degree

In Sect. 4.2 we will show that every function f can be perfectly encoded by a degree 3 encoding in $\mathbf{NC}_4^0$ whose complexity is polynomial in the size of the branching program that computes f. In [92, Corollary 5.9] it was shown that most boolean functions do not admit perfectly-private randomized encoding of degree 2 regardless of the efficiency of the encoding. (In fact, an exact characterization of the class of functions that admit such an encoding is given.) Hence, 3 is the minimal degree which is sufficient to (perfectly) encode all functions. We now show that locality 4 is also minimal with respect to perfect encoding.

---

[5]In fact, this is a specific instance of the statistical difference problem which was shown to be complete for the class SZK [129].

We say that a function $f : \{0, 1\}^n \to \{0, 1\}^l$ is ε-balanced if $\| f(U_n) - U_l \| \leq \varepsilon$. When $\varepsilon = 0$ we will say that f is balanced. The following claim shows that if a c-local function is $\frac{1}{2} \cdot 2^{-c}$-balanced then it is actually balanced and its degree is bounded by $c - 1$.

Claim 3.1 *Let* $f : \{0, 1\}^c \to \{0, 1\}$ *be a* $\frac{1}{2} \cdot 2^{-c}$*-balanced function. Then,* f *is balanced and its degree is at most* $c - 1$.

Proof In the following we will identify the truth table of a function $g : \{0, 1\}^c \to \{0, 1\}$ with a string $z \in \{0, 1\}^{2^c}$ which is indexed by c-bit strings such that $z_x = g(x)$.

First note that when f is not balanced its bias $| \Pr_x[f(x) = 1] - \frac{1}{2} |$ is at least 2^{-c}. Hence, in this case $\| f(U_n) - U_1 \| \geq 2^{-c}$ and so it cannot be $\frac{1}{2} \cdot 2^{-c}$-balanced. We conclude that f is balanced.

Now assume, towards a contradiction, that f is a function of degree exactly c. Then we can write f as a sum of monomials $f(x) = T_1(x) + \cdots + T_k(x)$ where $T_1(x) = x_1 \cdots x_c$ and $T_2, \ldots, T_k$ are monomials of degree $\leq c - 1$. Let z be the truth table of f, and let $z^{(i)}$ be the truth table of T_i. The following three claims show that the truth table z of f has an odd number of ones and thus f cannot be balanced: (1) $z = z^{(1)} \oplus \cdots \oplus z^{(k)}$; (2) the number of ones in $z^{(1)}$ is odd; and (3) for $1 < i \leq k$, the number of ones in $z^{(i)}$ is even. The first two claims can be easily verified, and claim (3) follows from the fact that each T_i depends on a strict subset of the inputs. $\square$

We also need the following claim.

Claim 3.2 *Let* $f : \{0, 1\}^n \to \{0, 1\}^l$ *be an* ε*-balanced function which is perfectly encoded by* $\hat{f} : \{0, 1\}^n \times \{0, 1\}^m \to \{0, 1\}^s$. *Then,* $\hat{f}$ *is also* ε*-balanced.*

Proof Let S be the balanced simulator of $\hat{f}$. Then,

$$
\begin{aligned}
\| \hat{f}(U_n, U_m) - U_s \| &\leq \| \hat{f}(U_n, U_m) - S(f(U_n)) \| + \| S(f(U_n)) - S(U_l) \| \\
&\quad + \| S(U_l) - U_s \| \\
&= 0 + \| S(f(U_n)) - S(U_l) \| + 0 \\
&\leq \| f(U_n) - U_l \| \leq \varepsilon,
\end{aligned}
$$

where the first inequality follows from Fact 2.1, the first equality is due to perfect privacy, Fact 2.5 and the fact that S is a balanced simulator, the second inequality follows from Fact 2.3, and the last inequality is due to the fact that f is ε-balanced. $\square$

By combining the above claims and the results of [92, Corollary 5.9], we get the following corollary.

Corollary 3.1 *For sufficiently large* n, *most boolean functions* $f : \{0, 1\}^n \to \{0, 1\}$ *cannot be perfectly encoded by functions* $\hat{f} \in \mathbf{NC}_3^0$.

Proof We begin by proving the claim for a function f which is 2^{-4}-balanced and cannot be perfectly encoded by a degree 2 encoding. Assume, towards a contradiction, that f is perfectly encoded by a 3-local function $\hat{f}$. Then, by Claim 3.2, $\hat{f}$ is also 2^{-4}-balanced and thus, by Claim 3.1, $\hat{f}$ is a degree-2 encoding and we derive a contradiction.

We complete the proof by noting that most functions satisfy the above properties. Indeed, for sufficiently large n, all but a negligible fraction of the boolean functions over n-bits are 2^{-4}-balanced. (This can be proved by choosing a random function and applying a Chernoff bound.) Also, by [92, Corollary 5.9], only a negligible fraction of the boolean functions can be perfectly encoded by degree 2 functions. $\square$

In particular, it follows from [92, Corollary 5.9] that the function $x_1 \cdot x_2 \cdot x_3 + x_4$ cannot be perfectly encoded by a degree 2 function, hence, by Corollary 3.1, this function cannot be perfectly encoded by a 3-local function (regardless of the complexity of the encoding).

Chapter 4
Cryptography in NC⁰

Abstract In this chapter we show that randomized encoding preserves the security of many cryptographic primitives. We also construct an (information-theoretic) encoding in $\mathbf{NC}_4^0$ for any function in $\mathbf{NC}^1$ or even $\oplus \mathbf{L}/poly$. The combination of these results gives a compiler that takes as an input a code of an $\mathbf{NC}^1$ implementation of some cryptographic primitive and generates an $\mathbf{NC}_4^0$ implementation of the same primitive.

4.1 Introduction

As indicated in Chap. 1, the possibility of implementing most cryptographic primitives in $\mathbf{NC}^0$ was left wide open. We present a positive answer to this basic question, showing that surprisingly many cryptographic tasks can be performed in constant parallel time. Since the existence of cryptographic primitives implies that $\mathbf{P} \neq \mathbf{NP}$, we cannot expect unconditional results and have to rely on some unproven assumptions.[1] However, we avoid relying on *specific* intractability assumptions. Instead, we assume the existence of cryptographic primitives in a relatively "high" complexity class and transform them to the seemingly degenerate complexity class $\mathbf{NC}^0$ without substantial loss of their cryptographic strength. These transformations are inherently non-black-box, thus providing further evidence for the usefulness of non-black-box techniques in cryptography.

4.1.1 Previous Work

Recall that a pseudorandom generator (PRG) $G : \{0,1\}^n \rightarrow \{0,1\}^l$ is a deterministic function that expands a short random n-bit string (aka "seed") into a longer "pseudorandom" string of length $l > n$. Pseudorandom *functions* [74] can viewed as PRGs with exponential output length and direct access. Linial et al. [108] show that

---

[1] This is not the case for non-cryptographic PRGs such as ε-biased generators, for which we do obtain unconditional results.

B. Applebaum, *Cryptography in Constant Parallel Time*,
Information Security and Cryptography,
DOI 10.1007/978-3-642-17367-7_4, © Springer-Verlag Berlin Heidelberg 2014

such functions cannot be computed even in $\mathbf{AC}^0$.[2] However, no such impossibility result is known for standard PRGs.

The existence of PRGs in $\mathbf{NC}^0$ has been studied in [50, 112]. Cryan and Miltersen [50] observe that there is no PRG in $\mathbf{NC}_2^0$, and prove that there is no PRG in $\mathbf{NC}_3^0$ achieving a superlinear stretch; namely, one that stretches n bits to $n + \omega(n)$ bits.[3] Mossel et al. [112] extend this impossibility to $\mathbf{NC}_4^0$. Viola [138] shows that a PRG in $\mathbf{AC}^0$ with superlinear stretch cannot be obtained from a one-way function (OWF) via non-adaptive black-box constructions. This result can be extended to rule out such a construction even if we start with a PRG whose stretch is sublinear. Negative results for other restricted computation models appear in [66, 147].

On the positive side, Impagliazzo and Naor [91] construct a (sublinear-stretch) PRG in $\mathbf{AC}^0$, relying on an intractability assumption related to the subset-sum problem. PRG candidates in $\mathbf{NC}^1$ (or even $\mathbf{TC}^0$) are more abundant, and can be based on a variety of standard cryptographic assumptions including ones related to the intractability of factoring [103, 117], discrete logarithms [36, 117, 144] and lattice problems [3, 86].[4]

Unlike the case of pseudorandom generators, the question of one-way functions in $\mathbf{NC}^0$ is relatively unexplored. The impossibility of OWFs in $\mathbf{NC}_2^0$ follows from the easiness of 2-SAT [50, 69]. Håstad [85] constructs a family of permutations in $\mathbf{NC}^0$ whose inverses are P-hard to compute. Cryan and Miltersen [50], improving on [1], present a circuit family in $\mathbf{NC}_3^0$ whose range decision problem is NP-complete. This, however, gives no evidence of cryptographic strength. Since any PRG is also a OWF, all PRG candidates cited above are also OWF candidates. (In fact, the one-wayness of an $\mathbf{NC}^1$ function often serves as the underlying cryptographic *assumption*.) Finally, Goldreich [69] suggests a candidate OWF in $\mathbf{NC}^0$, whose conjectured security does not follow from any well-known assumption.

4.1.2 Our Results

4.1.2.1 A General Compiler

Our main result is that any OWF (resp., PRG) in a relatively high complexity class, containing uniform $\mathbf{NC}^1$ and even $\oplus \mathbf{L}/poly$, can be efficiently "compiled" into a

---

[2]Indeed, the low noise-sensitivity of $\mathbf{AC}^0$ functions (cf. [137, Lemma 6.6]) allows us to efficiently distinguish them from truly random function, e.g., by querying the function on a pair of random points which are $1/\sqrt{n}$-close to each other in Hamming distance.

[3]From here on, we use a crude classification of PRGs into ones having sublinear, linear, or superlinear additive stretch. Note that a PRG stretching its seed by just one bit can be invoked *in parallel* (on seeds of length n^ε) to yield a PRG stretching its seed by $n^{1-\varepsilon}$ bits, for an arbitrary $\varepsilon > 0$. Also, an $\mathbf{NC}^0$ PRG with some linear stretch can be composed with itself a constant number of times to yield an $\mathbf{NC}^0$ PRG with an arbitrary linear stretch.

[4]In some of these constructions it seems necessary to allow a *collection* of $\mathbf{NC}^1$ PRGs, and use polynomial-time preprocessing to pick (once and for all) a random instance from this collection. This is similar to the more standard notion of OWF collection (cf. [70, Sect. 2.4.2]). See Appendix 4.9 for further discussion of this slightly relaxed notion of PRG.

corresponding OWF (resp., sublinear-stretch PRG) in $\mathbf{NC}_4^0$. The existence of OWF and PRG in this class is a mild assumption, implied in particular by most number-theoretic or algebraic intractability assumptions commonly used in cryptography. Hence, the existence of OWF and sublinear-stretch PRG in $\mathbf{NC}^0$ follows from a variety of standard assumptions and is not affected by the potential weakness of a particular algebraic structure. It is important to note that the PRG produced by our compiler will generally have a sublinear additive stretch even if the original PRG has a large stretch. However, one cannot do much better when insisting on an $\mathbf{NC}_4^0$ PRG, as there is no PRG with superlinear stretch in $\mathbf{NC}_4^0$ [112].

The above results extend to other cryptographic primitives including one-way permutation, encryption, signature, commitment, and collision-resistant hashing. Aiming at $\mathbf{NC}^0$ implementations, we can use our machinery in two different ways: (1) compile a primitive in a relatively high complexity class (say $\mathbf{NC}^1$) into its randomized encoding and show that the encoding inherits the security properties of this primitive; or (2) use known *reductions* between cryptographic primitives, together with $\mathbf{NC}^0$ primitives we already constructed (e.g., OWF or PRG), to obtain new $\mathbf{NC}^0$ primitives. Of course, this approach is useful only when the reduction itself is in $\mathbf{NC}^0$. If the reduction is in $\mathbf{NC}^1$ one can combine the two approaches: first apply the $\mathbf{NC}^1$ reduction to an $\mathbf{NC}^0$ primitive of type X that was already constructed (e.g., OWF or PRG) to obtain a new $\mathbf{NC}^1$ primitive of type Y, and then use the first approach to compile the latter primitive into an $\mathbf{NC}^0$ primitive (of type Y). As in the first approach, this construction requires us to prove that a randomized encoding of a primitive Y preserves its security. We mainly adopt the first approach, since most of the known reductions between primitives are not in $\mathbf{NC}^0$. (An exception in the case of symmetric encryption and zero-knowledge proofs will be discussed in Sect. 4.6.)

A Caveat It is important to note that in the case of two-party primitives (such as encryption schemes, signatures, commitments and zero-knowledge proofs) our compiler yields an $\mathbf{NC}^0$ sender (i.e., encrypting party, committing party, signer or prover, according to the case) but does not promise anything regarding the parallel complexity of the receiver (the decrypting party or verifier). In fact, we prove that, in all these cases, it is *impossible* to implement the receiver in $\mathbf{NC}^0$, regardless of the complexity of the sender. An interesting feature of the case of commitment is that we can also improve the parallel complexity at the receiver's end. Specifically, our receiver can be realized by an $\mathbf{NC}^0$ circuit with a single unbounded fan-in AND gate at the top. The same holds for applications of commitment such as coin-flipping and zero-knowledge proofs.

4.1.2.2 Parallel Reductions

In some cases our techniques yield $\mathbf{NC}^0$ reductions between different cryptographic primitives. (Unlike the results discussed above, here we consider *unconditional* reductions that do not rely on unproven assumptions.) Specifically, we get new $\mathbf{NC}^0$ constructions of PRG and non-interactive commitment from OWF and one-to-one

OWF, respectively. These reductions are obtained by taking a standard $\mathbf{NC}^1$ reduction of a primitive $\mathscr{P}$ from a primitive $\mathscr{P}'$, and applying our compiler to it. This works only for reductions that make non-adaptive calls to $\mathscr{P}'$. There are known reductions of PRGs and non-interactive commitments from OWF and one-to-one OWF which admit such a structure, and thus can be compiled into $\mathbf{NC}^0$ reductions. (We remark that if the original reduction uses the underlying primitive $\mathscr{P}'$ as a black-box then so does the new reduction. This should not be confused with the fact that the $\mathbf{NC}^0$ reduction is obtained by using the "code" of the original reduction. Indeed, all the $\mathbf{NC}^0$ reductions obtained in this chapter are black-box reductions.)

4.1.2.3 Non-cryptographic Generators

Our techniques can also be applied to obtain unconditional constructions of non-cryptographic PRGs. In particular, building on an ε-biased generator in $\mathbf{NC}^0_5$ constructed by Mossel et al. [112], we obtain a linear-stretch ε-biased generator in $\mathbf{NC}^0_3$. This generator has optimal locality, answering an open question posed in [112]. It is also essentially optimal with respect to stretch, since locality 3 does not allow for a superlinear stretch [50]. Our techniques apply also to other types of non-cryptographic PRGs such as generators for space-bounded computation [23, 119], yielding such generators (with sublinear stretch) in $\mathbf{NC}^0_3$.

4.1.3 Overview of Techniques

Our key observation is that instead of computing a given "cryptographic" function $f(x)$, it might suffice to compute a randomized encoding $\hat{f}(x, r)$ of f. Recall that $\hat{f}(x, r)$ has the following relation to f:

1. (Correctness.) For every fixed input x and a uniformly random choice of r, the output distribution $\hat{f}(x, r)$ forms a "randomized encoding" of $f(x)$, from which $f(x)$ can be decoded. That is, if $f(x) \neq f(x')$ then the random variables $\hat{f}(x, r)$ and $\hat{f}(x', r')$, induced by a uniform choice of r, r', should have disjoint supports.
2. (Privacy.) There exists a simulator algorithm that given $f(x)$ samples from the distribution $\hat{f}(x, r)$ induced by a uniform choice of r. That is, the distribution $\hat{f}(x, r)$ hides all the information about x except for the value $f(x)$. In particular, if $f(x) = f(x')$ then the random variables $\hat{f}(x, r)$ and $\hat{f}(x', r')$ should be identically distributed.

Each of these requirements alone can be satisfied by a trivial function $\hat{f}$ (e.g., $\hat{f}(x, r) = x$ and $\hat{f}(x, r) = 0$, respectively). However, the combination of the two requirements can be viewed as a non-trivial natural relaxation of the usual notion of computing. In a sense, the function $\hat{f}$ defines an "information-theoretically equivalent" representation of f.

For this approach to be useful in our context, two conditions should be met. First, we need to argue that a randomized encoding $\hat{f}$ can be *securely* used as a substitute for f. Second, we hope that this relaxation is sufficiently *liberal*, in the sense that it allows us to efficiently encode relatively complex functions f by functions $\hat{f}$ in $\mathbf{NC}^0$. These two issues are addressed in the following subsections.

4.1.3.1 Security of Randomized Encodings

To illustrate how a randomized encoding $\hat{f}$ can inherit the security features of f, consider the case where f is a OWF. We argue that the hardness of inverting $\hat{f}$ reduces to the hardness of inverting f. Indeed, a successful algorithm A for inverting $\hat{f}$ can be used to successfully invert f as follows: given an output y of f, apply the efficient sampling algorithm guaranteed by requirement 2 to obtain a random encoding $\hat{y}$ of y. Then, use A to obtain a preimage (x, r) of $\hat{y}$ under $\hat{f}$, and output x. It follows from requirement 1 that x is indeed a preimage of y under f. Moreover, if y is the image of a uniformly random x, then $\hat{y}$ is the image of a uniformly random pair (x, r). Hence, the success probability of inverting f is the same as that of inverting $\hat{f}$.

The above argument can tolerate some relaxations to the notion of randomized encoding. In particular, one can relax the second requirement to allow the distributions $\hat{f}(x, r)$ and $\hat{f}(x', r')$ be only *statistically close*, or even *computationally indistinguishable*. On the other hand, to maintain the security of other cryptographic primitives, it may be required to further strengthen this notion. For instance, when f is a PRG, the above requirements do not guarantee that the output of $\hat{f}$ is pseudorandom, or even that its output is longer than its input. However, by imposing suitable "regularity" requirements (e.g., the properties of *perfect* encoding cf. Definition 3.4) on the output encoding defined by $\hat{f}$, it can be guaranteed that if f is a PRG then so is $\hat{f}$. Thus, different security requirements suggest different variations of the above notion of randomized encoding.

4.1.3.2 Complexity of Randomized Encodings

It remains to address the second issue: can we encode a complex function f by an $\mathbf{NC}^0$ function $\hat{f}$? Our best solutions to this problem rely on the machinery of *randomizing polynomials,* described below. But first we outline a simple alternative approach[5] based on Barrington's theorem [24], combined with a randomization technique of Kilian [104].

Suppose f is a boolean function in $\mathbf{NC}^1$. (Non-boolean functions are handled by concatenation. See Lemma 3.1.) By Barrington's theorem, evaluating $f(x)$, for such a function f, reduces to computing an iterated product of polynomially many

---

[5]In fact, a modified version of this approach has been applied for constructing randomizing polynomials in [49].

elements $s_1, \ldots, s_m$ from the symmetric group S_5, where each s_i is determined by a single bit of x (i.e., for every i there exists j such that s_i is a function of x_j). Now, let $\hat{f}(x, r) = (s_1 r_1, r_1^{-1} s_2 r_2, \ldots, r_{m-2}^{-1} s_{m-1} r_{m-1}, r_{m-1}^{-1} s_m)$, where the random inputs r_i are picked uniformly and independently from S_5. It is not hard to verify that the output $(t_1, \ldots, t_m)$ of $\hat{f}$ is random subject to the constraint that $t_1 t_2 \cdots t_m = s_1 s_2 \cdots s_m$, where the latter product is in one-to-one correspondence to $f(x)$. It follows that $\hat{f}$ is a randomized encoding of f. Moreover, $\hat{f}$ has constant locality when viewed as a function over the alphabet S_5, and thus yields the qualitative result we are after.

However, the above construction falls short of providing a randomized encoding in $\mathbf{NC}^0$, since it is impossible to sample a uniform element of S_5 in $\mathbf{NC}^0$ (even up to a negligible statistical distance).[6] Also, this $\hat{f}$ does not satisfy the extra "regularity" properties required by more "sensitive" primitives such as PRGs or one-way permutations. The solutions presented next avoid these disadvantages and, at the same time, apply to a higher complexity class than $\mathbf{NC}^1$ and achieve a very small constant locality.

Randomizing Polynomials The concept of randomizing polynomials was introduced by Ishai and Kushilevitz [92] as a representation of functions by vectors of low-degree multivariate polynomials. (Interestingly, this concept was motivated by questions in the area of *information-theoretic* secure multiparty computation, which seems unrelated to the current context.) Randomizing polynomials capture the above encoding question within an algebraic framework. Specifically, a representation of $f(x)$ by randomizing polynomials is a randomized encoding $\hat{f}(x, r)$ as defined above, in which x and r are viewed as vectors over a finite field $\mathbb{F}$ and the outputs of $\hat{f}$ as multivariate polynomials in the variables x and r. In this work, we will always let $\mathbb{F} = \mathbb{F}_2$.

The most crucial parameter of a randomizing polynomials representation is its algebraic *degree*, defined as the maximal (total) degree of the outputs (i.e., the output multivariate polynomials) as a function of the input variables in x and r. (Note that both x and r count towards the degree.) Quite surprisingly, it is shown in [92, 93] that every boolean function $f : \{0, 1\}^n \to \{0, 1\}$ admits a representation by *degree-3* randomizing polynomials whose number of inputs and outputs is at most *quadratic* in its (mod-2) branching program size (cf. Sect. 2.3). (Moreover, this degree bound is tight in the sense that most boolean functions do not admit a degree-2 representation.) Recall that a representation of a non-boolean function can be obtained by concatenating representations of its output bits, using independent blocks of random inputs. (See Lemma 3.1.) This concatenation leaves the degree unchanged.

The above positive result implies that functions whose output bits can be computed in the complexity class $\oplus \mathbf{L}/poly$ admit an efficient representation by degree-3 randomizing polynomials. This also holds if one requires the most stringent notion of representation required by our applications (i.e., perfect encoding). We note, however, that different constructions from the literature [49, 92, 93] are incomparable

---

[6]Barrington's theorem generalizes to apply over arbitrary non-solvable groups. Unfortunately, there are no such groups whose order is a power of two.

in terms of their exact efficiency and the security-preserving features they satisfy. Hence, different constructions may be suitable for different applications.

Degree vs. Locality   Combining our general methodology with the above results on randomizing polynomials already brings us close to our goal, as it enables "degree-3 cryptography". Taking on from here, we show that any function $f : \{0, 1\}^n \to \{0, 1\}^m$ of algebraic degree d admits an efficient randomized encoding $\hat{f}$ of (degree d and) locality $d + 1$. That is, each output bit of $\hat{f}$ can be computed by a degree-d polynomial over $\mathbb{F}_2$ depending on at most $d + 1$ inputs and random inputs. Combined with the previous results, this allows us to make the final step from degree 3 to locality 4.

4.1.4 Organization

In Sect. 4.2 we construct a perfect and statistical encoding in $\mathbf{NC}_4^0$ for various log-space classes (i.e., $\oplus$L/*poly* and NL/*poly*). We then apply randomized encodings to obtain $\mathbf{NC}^0$ implementations of different primitives: OWFs (Sect. 4.3), cryptographic and non-cryptographic PRGs (Sect. 4.4), collision-resistant hashing (Sect. 4.5), public-key and symmetric encryption (Sect. 4.6) and other cryptographic primitives (Sect. 4.7). We conclude in Sect. 4.8 with a summary of the results obtained in this chapter.

4.2   NC^0 Encoding for Functions in $\oplus$L/*poly* and NL/*poly*

In this section we construct randomized encodings in $\mathbf{NC}^0$. We first review a construction from [93] of degree-3 randomizing polynomials based on mod-2 branching programs and analyze some of its properties. Next, we introduce a general locality reduction technique, allowing us to transform a degree-d encoding to a $(d + 1)$-local encoding. Finally, we discuss extensions to other types of BPs.

4.2.1 Degree-3 Randomizing Polynomials from mod-2 BPs

Let $BP = (G, \phi, s, t)$ be a mod-2 BP of size ℓ, computing a boolean[7] function $f : \{0, 1\}^n \to \{0, 1\}$; that is, $f(x) = 1$ if and only if the number of paths from s to t in G_x equals 1 modulo 2. Fix some topological ordering of the vertices of G, where the source vertex s is labeled 1 and the terminal vertex t is labeled ℓ. Let $A(x)$ be the

---

[7] The following construction generalizes naturally to a (counting) mod-p BP, computing a function $f : \{0, 1\}^n \to Z_p$. In this work, however, we will only be interested in the case $p = 2$.

$$\begin{pmatrix} 1 & r_1^{(1)} & r_2^{(1)} & \cdot & \cdot & r_{\ell-2}^{(1)} \\ 0 & 1 & \cdot & \cdot & \cdot & \cdot \\ 0 & 0 & 1 & \cdot & \cdot & \cdot \\ 0 & 0 & 0 & 1 & \cdot & \cdot \\ 0 & 0 & 0 & 0 & 1 & r_{\binom{\ell-1}{2}}^{(1)} \\ 0 & 0 & 0 & 0 & 0 & 1 \end{pmatrix} \begin{pmatrix} * & * & * & * & * & * \\ -1 & * & * & * & * & * \\ 0 & -1 & * & * & * & * \\ 0 & 0 & -1 & * & * & * \\ 0 & 0 & 0 & -1 & * & * \\ 0 & 0 & 0 & 0 & -1 & * \end{pmatrix} \begin{pmatrix} 1 & 0 & 0 & 0 & 0 & r_1^{(2)} \\ 0 & 1 & 0 & 0 & 0 & r_2^{(2)} \\ 0 & 0 & 1 & 0 & 0 & \cdot \\ 0 & 0 & 0 & 1 & 0 & \cdot \\ 0 & 0 & 0 & 0 & 1 & r_{\ell-2}^{(2)} \\ 0 & 0 & 0 & 0 & 0 & 1 \end{pmatrix}$$

Fig. 4.1 The matrices $R_1(r^{(1)})$, $L(x)$ and $R_2(r^{(2)})$ (from left to right). The symbol $*$ represents a degree-1 polynomial in an input variable

$\ell \times \ell$ adjacency matrix of G_x viewed as a formal matrix whose entries are degree-1 polynomials in the input variables x. Specifically, the (i, j) entry of $A(x)$ contains the value of $\phi(i, j)$ on x if (i, j) is an edge in G, and 0 otherwise. (Hence, $A(x)$ contains the constant 0 on and below the main diagonal, and degree-1 polynomials in the input variables above the main diagonal.) Define $L(x)$ as the submatrix of $A(x) - I$ obtained by deleting column s and row t (i.e., the first column and the last row). As before, each entry of $L(x)$ is a degree-1 polynomial in a single input variable x_i; moreover, $L(x)$ contains the constant -1 in each entry of its second diagonal (the one below the main diagonal) and the constant 0 below this diagonal. (See Fig. 4.1.)

Fact 4.1 ([93]) $f(x) = \det(L(x))$, where the determinant is computed over $\mathbb{F}_2$.

Proof Sketch Since G is acyclic, the number of $s - t$ paths in G_x mod 2 can be written as $(I + A(x) + A(x)^2 + \cdots + A(x)^\ell)_{s,t} = (I - A(x))_{s,t}^{-1}$ where I denotes an $\ell \times \ell$ identity matrix and all arithmetic is over $\mathbb{F}_2$. Recall that $L(x)$ is the submatrix of $A(x) - I$ obtained by deleting column s and row t. Hence, expressing $(I - A(x))_{s,t}^{-1}$ using the corresponding cofactor of $I - A(x)$, we have:

$$(I - A(x))_{s,t}^{-1} = (-1)^{s+t} \frac{\det(-L(x))}{\det(I - A(x))} = \det L(x). \qquad \square$$

Let $r^{(1)}$ and $r^{(2)}$ be vectors over $\mathbb{F}_2$ of length $\sum_{i=1}^{\ell-2} i = \binom{\ell-1}{2}$ and $\ell - 2$, respectively. Let $R_1(r^{(1)})$ be an $(\ell - 1) \times (\ell - 1)$ matrix with 1's on the main diagonal, 0's below it, and $r^{(1)}$'s elements in the remaining $\binom{\ell-1}{2}$ entries above the diagonal (a unique element of $r^{(1)}$ is assigned to each matrix entry). Let $R_2(r^{(2)})$ be an $(\ell - 1) \times (\ell - 1)$ matrix with 1's on the main diagonal, $r^{(2)}$'s elements in the rightmost column, and 0's in each of the remaining entries. (See Fig. 4.1.)

Fact 4.2 ([93]) Let M, M' be $(\ell - 1) \times (\ell - 1)$ matrices that contain the constant -1 in each entry of their second diagonal and the constant 0 below this diagonal. Then, $\det(M) = \det(M')$ if and only if there exist $r^{(1)}$ and $r^{(2)}$ such that $R_1(r^{(1)}) M R_2(r^{(2)}) = M'$.

Proof Sketch Suppose that $R_1(r^{(1)})MR_2(r^{(2)}) = M'$ for some $r^{(1)}$ and $r^{(2)}$. Then, since $\det(R_1(r^{(1)})) = \det(R_2(r^{(2)})) = 1$, it follows that $\det(M) = \det(M')$.

For the second direction assume that $\det(M) = \det(M')$. We show that there exist $r^{(1)}$ and $r^{(2)}$ such that $R_1(r^{(1)})MR_2(r^{(2)}) = M'$. Multiplying M by a matrix $R_1(r^{(1)})$ on the left is equivalent to adding to each row of M a linear combination of the rows below it. On the other hand, multiplying M by a matrix $R_2(r^{(2)})$ on the right is equivalent to adding to the last column of M a linear combination of the other columns. Observe that a matrix M that contains the constant -1 in each entry of its second diagonal and the constant 0 below this diagonal can be transformed, using such left and right multiplications, to a canonical matrix H_y containing -1's in its second diagonal, an arbitrary value y in its top-right entry, and 0's elsewhere. Since $\det(R_1(r^{(1)})) = \det(R_2(r^{(2)})) = 1$, we have $\det(M) = \det(H_y) = y$. Thus, when $\det(M) = \det(M') = y$ we can write $H_y = R_1(r^{(1)})MR_2(r^{(2)}) = R_1(s^{(1)})M'R_2(s^{(2)})$ for some $r^{(1)}, r^{(2)}, s^{(1)}, s^{(2)}$. Multiplying both sides by $R_1(s^{(1)})^{-1}, R_2(s^{(2)})^{-1}$, and observing that each set of matrices $R_1(\cdot)$ and $R_2(\cdot)$ forms a multiplicative group finishes the proof. $\qquad\square$

Lemma 4.1 (Implicit in [93]) *Let BP be a mod-2 branching program computing the boolean function f. Define a degree-3 function $\hat{f}(x, (r^{(1)}, r^{(2)}))$ whose outputs contain the $\binom{\ell}{2}$ entries on or above the main diagonal of the matrix $R_1(r^{(1)})L(x)R_2(r^{(2)})$. Then, $\hat{f}$ is a perfect randomized encoding of f.*

Proof We start by showing that the encoding is stretch preserving. The length of the random input of $\hat{f}$ is $m = \binom{\ell-1}{2} + \ell - 2 = \binom{\ell}{2} - 1$ and its output length is $s = \binom{\ell}{2}$. Thus we have $s = m + 1$, and since f is a boolean function its encoding $\hat{f}$ preserves its stretch.

We now describe the decoder and the simulator. Given an output of $\hat{f}$, representing a matrix M, the decoder B simply outputs $\det(M)$. (Note that the entries below the main diagonal of this matrix are constants and therefore are not included in the output of $\hat{f}$.) By Facts 4.1 and 4.2, $\det(M) = \det(L(x)) = f(x)$, hence the decoder is perfect.

The simulator S, on input $y \in \{0, 1\}$, outputs the $\binom{\ell}{2}$ entries on and above the main diagonal of the matrix $R_1(r^{(1)})H_y R_2(r^{(2)})$, where $r^{(1)}, r^{(2)}$ are randomly chosen, and H_y is the $(\ell - 1) \times (\ell - 1)$ matrix that contains -1's in its second diagonal, y in its top-right entry, and 0's elsewhere.

By Facts 4.1 and 4.2, for every $x \in \{0, 1\}^n$ the supports of $\hat{f}(x, U_m)$ and of $S(f(x))$ are equal. Specifically, these supports include all strings in $\{0, 1\}^s$ representing matrices with determinant $f(x)$. Since the supports of $S(0)$ and $S(1)$ form a disjoint partition of the entire space $\{0, 1\}^s$ (by Fact 4.2) and since S uses $m = s - 1$ random bits, it follows that $|\text{support}(S(b))| = 2^m$, for $b \in \{0, 1\}$. Since both the simulator and the encoding use m random bits, it follows that both distributions, $\hat{f}(x, U_m)$ and $S(f(x))$, are uniform over their support and therefore are equivalent. Finally, since the supports of $S(0)$ and $S(1)$ halve the range of $\hat{f}$ (that is, $\{0, 1\}^s$), the simulator is also balanced. $\qquad\square$

4.2.2   Reducing the Locality

It remains to convert the degree-3 encoding into one in **NC**^{0}. To this end, we show how to construct for any degree-d function (where d is constant) a $(d+1)$-local perfect encoding. Using the composition lemma, we can obtain an **NC**^{0} encoding of a function by first encoding it as a constant-degree function, and then applying the locality construction.

The idea for the locality construction is to represent a degree-d polynomial as a sum of monomials, each having locality d, and randomize this sum using a variant of the method for randomizing group product, described in Sect. 4.1.3.2. (A direct use of the latter method over the group Z_2 gives a $(d+2)$-local encoding instead of the $(d+1)$-local one obtained here.)

Construction 4.1 (Locality construction)   *Let* $f(x) = T_1(x) + \cdots + T_k(x)$, *where* $f, T_1, \ldots, T_k : \mathbb{F}_2^n \to \mathbb{F}_2$ *and summation is over* $\mathbb{F}_2$. *The local encoding* $\hat{f} : \mathbb{F}_2^{n+(2k-1)} \to \mathbb{F}_2^{2k}$ *is defined by:*

$$\hat{f}\left(x, \left(r_1, \ldots, r_k, r_1', \ldots, r_{k-1}'\right)\right)$$
$$\stackrel{\text{def}}{=} \left(T_1(x) - r_1, T_2(x) - r_2, \ldots, T_k(x) - r_k, \right.$$
$$\left. r_1 - r_1', r_1' + r_2 - r_2', \ldots, r_{k-2}' + r_{k-1} - r_{k-1}', r_{k-1}' + r_k\right).$$

For example, applying the locality construction to the polynomial $x_1 x_2 + x_2 x_3 + x_4$ results in the encoding $(x_1 x_2 - r_1, x_2 x_3 - r_2, x_4 - r_3, r_1 - r_1', r_1' + r_2 - r_2', r_2' + r_3)$.

Lemma 4.2 (Locality lemma)   *Let* f *and* $\hat{f}$ *be as in Construction* 4.1. *Then,* $\hat{f}$ *is a perfect randomized encoding of* f. *In particular, if* f *is a degree-d polynomial written as a sum of monomials, then* $\hat{f}$ *is a perfect encoding of* f *with degree d and locality* $\max(d+1, 3)$.

Proof   Since $m = 2k - 1$ and $s = 2k$, the encoding $\hat{f}$ is stretch preserving. Moreover, given $\hat{y} = \hat{f}(x, r)$ we can decode the value of $f(x)$ by summing up the bits of $\hat{y}$. It is not hard to verify that such a decoder never errs. To prove perfect privacy we define a simulator as follows. Given $y \in \{0, 1\}$, the simulator S uniformly chooses $2k - 1$ random bits $r_1, \ldots, r_{2k-1}$ and outputs $(r_1, \ldots, r_{2k-1}, y - (r_1 + \cdots + r_{2k-1}))$. Obviously, $S(y)$ is uniformly distributed over the $2k$-length strings whose bits sum up to y over $\mathbb{F}_2$. It thus suffices to show that the outputs of $\hat{f}(x, U_m)$ are uniformly distributed subject to the constraint that they add up to $f(x)$. This follows by observing that, for any x and any assignment $w \in \{0, 1\}^{2k-1}$ to the first $2k - 1$ outputs of $\hat{f}(x, U_m)$, there is a unique way to set the random inputs r_i, r_i' so that the output of $\hat{f}(x, (r, r'))$ is consistent with w. Indeed, for $1 \le i \le k$, the values of x, w_i uniquely determine r_i. For $1 \le i \le k - 1$, the values w_{k+i}, r_i, r_{i-1}' determine r_i' (where $r_0' \stackrel{\text{def}}{=} 0$). Therefore, $S(f(x)) \equiv \hat{f}(x, U_m)$. Moreover, S is balanced since the supports of $S(0)$ and $S(1)$ halve $\{0, 1\}^s$ and $S(y)$ is uniformly distributed over its support for $y \in \{0, 1\}$. $\square$

In Sect. 4.2.3 we describe a graph-based generalization of Construction 4.1, which in some cases can give rise to a (slightly) more compact encoding $\hat{f}$.

We now present the main theorem of this section.

Theorem 4.1 $\oplus$L/*poly* $\subseteq$ **PREN**. *Moreover, any* $f \in \oplus$L/*poly admits a perfect randomized encoding in* $\mathbf{NC}_4^0$ *whose degree is* 3.

Proof The theorem is derived by combining the degree-3 construction of Lemma 4.1 together with the Locality Lemma 4.2, using the Composition Lemma 3.3 and the Concatenation Lemma 3.2. $\square$

Remark 4.1 An alternative construction of perfect randomized encodings in $\mathbf{NC}^0$ can be obtained using a randomizing polynomials construction from [93, Sect. 3], which is based on an information-theoretic variant of Yao's garbled circuit technique [145]. This construction yields an encoding with a (large) constant locality, without requiring an additional "locality reduction" step (of Construction 4.1). This construction is weaker than the current one in that it only efficiently applies to functions in $\mathbf{NC}^1$ rather than $\oplus$L/*poly*. For functions in $\mathbf{NC}^1$, the complexity of this alternative (in terms of randomness and output length) is incomparable to the complexity of the current construction.

There are variants of the above construction that can handle non-deterministic branching programs as well, at the expense of losing perfectness [92, 93]. In particular, Theorem 2 in [93] encodes non-deterministic branching programs by perfectly-correct statistically-private functions of degree 3 (over $\mathbb{F}_2$). Hence, by using Lemmas 4.2, 3.3, we get a perfectly-correct statistically-private encoding in $\mathbf{NC}_4^0$ for functions in NL/*poly*. (In fact, we can also use [92, 93] to obtain *perfectly-private* statistically-correct encoding in $\mathbf{NC}_4^0$ for non-deterministic branching programs.) Based on the above, we get the following theorem:

Theorem 4.2 NL/*poly* $\subseteq$ **SREN**. *Moreover, any* $f \in$ NL/*poly admits a perfectly-correct statistically-private randomized encoding in* $\mathbf{NC}_4^0$.

We can rely on Theorem 4.1 to derive the following corollary.

Corollary 4.1 *Any function* f *in* **PREN** (*resp.*, **SREN**, **CREN**) *admits a perfect* (*resp., statistical, computational*) *randomized encoding of degree* 3 *and locality* 4 (*i.e., in* $\mathbf{NC}_4^0$).

Proof We first encode f by a perfect (resp., statistical, computational) encoding $\hat{f}$ in $\mathbf{NC}^0$, guaranteed by the fact that f is in **PREN** (resp., **SREN**, **CREN**). Then, since $\hat{f}$ is in $\oplus$L/*poly*, we can use Theorem 4.1 to perfectly encode $\hat{f}$ by a function

$\hat{f}'$ in **NC**$^0_4$ whose degree is 3. By the Composition Lemmas 3.3 and 3.4, $\hat{f}'$ perfectly (resp. statistically, computationally) encodes the function f. $\square$

4.2.3 A Generalization of the Locality Construction

In the Locality Construction 4.1, we showed how to encode a degree d function by an **NC**$^0_{d+1}$ encoding. We now describe a graph based construction that generalizes the previous one. The basic idea is to view the encoding $\hat{f}$ as a graph. The nodes of the graph are labeled by terms of f and the edges by random inputs of $\hat{f}$. With each node we associate an output of $\hat{f}$ in which we add to its term the labels of the edges incident to the node. Formally,

Construction 4.2 (General locality construction) *Let* $f(x) = T_1(x) + \cdots + T_k(x)$, *where* $f, T_1, \ldots, T_k : \mathbb{F}_2^n \to \mathbb{F}_2$ *and summation is over* $\mathbb{F}_2$. *Let* $G = (V, E)$ *be a directed graph with* k *nodes* $V = \{1, \ldots, k\}$ *and* m *edges. The encoding* $\hat{f}_G : \mathbb{F}_2^{n+m} \to \mathbb{F}_2^k$ *is defined by:*

$$\hat{f}_G\left(x, (r_{i,j})_{(i,j)\in E}\right) \stackrel{\text{def}}{=} \left(T_i(x) + \sum_{j|(j,i)\in E} r_{j,i} - \sum_{j|(i,j)\in E} r_{i,j} \right)_{i=1}^k .$$

From here on, we will identify with the directed graph G its underlying undirected graph. The above construction yields a perfect encoding when G is a tree (see Lemma 4.3 below). The locality of an output bit of $\hat{f}_G$ is the locality of the corresponding term plus the degree of the node in the graph. The locality construction described in Construction 4.1 attempts to minimize the maximal locality of a node in the graph; hence it adds k "dummy" 0 terms to f and obtains a tree in which all of the k non-dummy terms of f are leaves, and the degree of each dummy term is at most 3. When the terms of f vary in their locality, a more compact encoding $\hat{f}$ can be obtained by increasing the degree of nodes which represent terms with lower locality.

Lemma 4.3 (Generalized locality lemma) *Let* f *and* $\hat{f}_G$ *be as in Construction 4.2. Then,*

1. $\hat{f}_G$ *is a perfectly correct encoding of* f.
2. *If* G *is connected, then* $\hat{f}_G$ *is also a balanced encoding of* f *(and in particular it is perfectly private).*
3. *If* G *is a tree, then* $\hat{f}_G$ *is also stretch preserving; that is,* $\hat{f}_G$ *perfectly encodes* f.

Proof (1) Given $\hat{y} = \hat{f}_G(x, r)$ we decode $f(x)$ by summing up the bits of $\hat{y}$. Since each random variable $r_{i,j}$ appears only in the i-th and j-th output bits, it contributes 0 to the overall sum and therefore the bits of $\hat{y}$ always add up to $f(x)$.

To prove (2) we use the same simulator as in the locality construction (see proof of Lemma 4.2). Namely, given $y \in \{0, 1\}$, the simulator S chooses $k - 1$ random bits $r_1, \ldots, r_{k-1}$ and outputs $(r_1, \ldots, r_{k-1}, y - (r_1 + \cdots + r_{k-1}))$. This simulator is balanced since the supports of $S(0)$ and $S(1)$ halve $\{0, 1\}^k$ and $S(y)$ is uniformly distributed over its support for $y \in \{0, 1\}$. We now prove that $\hat{f}_G(x, U_m) \equiv S(f(x))$. Since the support of $S(f(x))$ contains exactly 2^{k-1} strings (namely, all k-bit strings whose bits sum up to $f(x)$), it suffices to show that for any input x and output $w \in \mathrm{support}(S(f(x)))$ there are $2^m / 2^{k-1}$ random inputs r such that $\hat{f}_G(x, r) = w$. (Note that $m \geq k - 1$ since G is connected.) Let $T \subseteq E$ be a spanning tree of G. We argue that for any assignment to the $m - (k - 1)$ random variables that correspond to edges in $E \setminus T$ there exists an assignment to the other random variables that is consistent with w and x. Fix some assignment to the edges in $E \setminus T$. We now recursively assign values to the remaining edges. In each step we make sure that some leaf is consistent with w by assigning the corresponding value to the edge connecting this leaf to the graph. Then, we prune this leaf and repeat the above procedure. Formally, let i be a leaf which is connected to T by an edge $e \in T$. Assume, without loss of generality, that e is an incoming edge for i. We set r_e to $w_i - (T_i(x) + \sum_{j|(j,i)\in E\setminus T} r_{j,i} - \sum_{j|(i,j)\in E\setminus T} r_{i,j})$, and remove i from T. By this we ensure that the i-th bit of $\hat{f}_G(x, r)$ is equal to w_i. (This equality will not be violated by the following steps as i is removed from T.) We continue with the above step until the tree consists of one node. Since the outputs of $\hat{f}_G(x, r)$ always sum up to $f(x)$ it follows that this last bit of $\hat{f}_G(x, r)$ is equal to the corresponding bit of w. Thus, there are at least $2^{|E\setminus T|} = 2^{m-(k-1)}$ values of r that lead to w as required.

Finally, to prove (3) note that when G is a tree we have $m = k - 1$, and therefore the encoding is stretch preserving; combined with (1) and (2) $\hat{f}_G$ is also perfect. □

4.3 One-Way Functions in $\mathbf{NC}^0$

A *one-way function* (OWF) $f : \{0, 1\}^* \to \{0, 1\}^*$ is a polynomial-time computable function that is hard to invert; namely, every (non-uniform) polynomial time algorithm that tries to invert f on input $f(x)$, where x is picked from U_n, succeeds only with a negligible probability. Formally,

Definition 4.1 (One-way function) A function $f : \{0, 1\}^* \to \{0, 1\}^*$ is called a *one-way function* (OWF) if it satisfies the following two properties:

- **Easy to compute**. There exists a deterministic polynomial-time algorithm computing $f(x)$.
- **Hard to invert**. For every non-uniform polynomial-time algorithm, A, we have

$$\Pr_{x \leftarrow U_n} \left[A\left(1^n, f(x)\right) \in f^{-1}\left(f(x)\right) \right] \leq \mathrm{neg}(n).$$

The function f is called *weakly one-way* if the second requirement is replaced with the following (weaker) one:

- **Slightly hard to invert**. There exists a polynomial $p(\cdot)$, such that for every (non-uniform) polynomial-time algorithm, A, and all sufficiently large n's

$$\Pr_{x \leftarrow U_n}\left[A\left(1^n, f(x)\right) \notin f^{-1}\left(f(x)\right)\right] > \frac{1}{p(n)}.$$

The above definition naturally extends to functions whose domain is restricted to some infinite subset $I \subset \mathbb{N}$ of the possible input lengths, such as ones defined by a randomized encoding $\hat{f}$. As argued in Remark 3.1, such a partially defined OWF can be augmented into a fully defined OWF provided that the set I is polynomially-dense and efficiently recognizable (which is a feature of functions $\hat{f}$ obtained via a uniform encodings).

4.3.1 Key Lemmas

In the following we show that a perfectly correct and statistically(or even computationally) private randomized encoding $\hat{f}$ of a OWF f is also a OWF. The idea, as described in Sect. 4.1.3.1, is to argue that the hardness of inverting $\hat{f}$ reduces to the hardness of inverting f. The case of a randomized encoding that does not enjoy perfect correctness is more involved and will be dealt with later in this section.

Lemma 4.4 *Suppose that $f : \{0, 1\}^* \to \{0, 1\}^*$ is a one-way function (respectively, weak one-way function) and $\hat{f}(x, r)$ is a perfectly correct, computationally private encoding of f. Then $\hat{f}$, viewed as a single-argument function, is also one-way (respectively, weakly one-way).*

Proof Let $s = s(n), m = m(n)$ be the lengths of the output and of the random input of $\hat{f}$ respectively. Note that $\hat{f}$ is defined on input lengths of the form $n + m(n)$; we prove that it is hard to invert on these inputs. Let $\hat{A}$ be an adversary that inverts $\hat{f}(x, r)$ with success probability $\phi(n+m)$. We use $\hat{A}$ to construct an efficient adversary A that inverts f with similar success. On input $(1^n, y)$, the adversary A runs S, the simulator of $\hat{f}$, on the input $(1^n, y)$ and gets a string $\hat{y}$ as the output of S. Next, A runs the inverter $\hat{A}$ on the input $(1^{n+m}, \hat{y})$, getting (x', r') as the output of $\hat{A}$ (i.e., $\hat{A}$ "claims" that $\hat{f}(x', r') = \hat{y}$). A terminates with output x'.

COMPLEXITY: Since S and $\hat{A}$ are both polynomial-time algorithms, and since $m(n)$ is polynomially bounded, it follows that A is also a polynomial-time algorithm.

CORRECTNESS: We analyze the success probability of A on input $(1^n, f(x))$ where $x \leftarrow U_n$. Let us assume for a moment that the simulator S is perfect. Observe that, by perfect correctness, if $f(x) \neq f(x')$ then the support sets of $\hat{f}(x, U_m)$ and $\hat{f}(x', U_m)$ are disjoint. Moreover, by perfect privacy the string $\hat{y}$, generated by $\hat{A}$, is always in the support of $\hat{f}(x, U_m)$. Hence, if $\hat{A}$ succeeds (that is, indeed $\hat{y} = \hat{f}(x', r')$) then so does A (namely, $f(x') = y$). Finally, observe that (by Fact 2.5) the input $\hat{y}$ on which A invokes $\hat{A}$ is distributed identically to $\hat{f}_n(U_n, U_{m(n)})$, and therefore A succeeds with probability $\geq \phi(n+m)$. Formally, we can write,

$$\Pr_{x \leftarrow U_n}\left[A\left(1^n, f(x)\right) \in f^{-1}\left(f(x)\right)\right]$$

$$\geq \Pr_{x \leftarrow U_n, \hat{y} \leftarrow S(1^n, f(x))}\left[\hat{A}\left(1^{n+m}, \hat{y}\right) \in \hat{f}^{-1}(\hat{y})\right]$$

$$= \Pr_{x \leftarrow U_n, r \leftarrow U_{m(n)}}\left[\hat{A}\left(1^{n+m}, \hat{f}_n(x, r)\right) \in \hat{f}^{-1}\left(\hat{f}(x, r)\right)\right]$$

$$\geq \phi(n+m).$$

We now show that when S is computationally private, we lose only negligible success probability in the above; that is, A succeeds with probability $\geq \phi(n+m) - \text{neg}(n)$. To see this, it will be convenient to define a distinguisher D that on input $(1^n, y, \hat{y})$ computes $(x', r') = \hat{A}(1^{n+m}, \hat{y})$, and outputs 1 if $f(x') = y$ and 0 otherwise. Clearly, the success probability of A on $f(U_n)$ can be written as $\Pr_{x \leftarrow U_n}[D(1^n, f(x), S(f(x))) = 1]$. On the other hand, we showed above that $\Pr_{x \leftarrow U_n, r \leftarrow U_{m(n)}}[D(1^n, f(x), \hat{f}(x, r)) = 1] \geq \phi(n+m)$. Also, by the computational privacy of S, and Facts 2.10, 2.8, we have

$$\left(f(U_n), S\left(f(U_n)\right)\right) \stackrel{c}{\equiv} \left(f(U_n), \hat{f}(U_n, U_{m(n)})\right).$$

Hence, since D is polynomial-time computable, we have

$$\Pr_{x \leftarrow U_n}\left[D\left(1^n, f(x), S\left(f(x)\right)\right) = 1\right]$$

$$\geq \Pr_{x \leftarrow U_n, r \leftarrow U_{m(n)}}\left[D\left(1^n, f(x), \hat{f}(x, r)\right) = 1\right] - \text{neg}(n)$$

$$\geq \phi(n+m) - \text{neg}(n).$$

Since f is hard to invert (respectively, slightly hard to invert), the quantity $\phi(n+m)$ must be negligible (respectively, bounded by $1 - 1/p(n)$ for some fixed polynomial $p(\cdot)$) and so $\hat{f}$ is also hard to invert (respectively, slightly hard to invert). $\qquad\square$

The efficiency of the simulator S is essential for Lemma 4.4 to hold. Indeed, without this requirement one could encode any one-way permutation f by the identity function $\hat{f}(x) = x$, which is obviously not one-way. (Note that the output of $\hat{f}(x)$ can be simulated inefficiently based on $f(x)$ by inverting f.)

The perfect correctness requirement is also essential for Lemma 4.4 to hold. To see this, consider the following example. Suppose f is a one-way permutation. Consider the encoding $\hat{f}(x, r)$ which equals $f(x)$ except if r is the all-zero string, in which case $\hat{f}(x, r) = x$. This is a statistically-correct and statistically-private encoding, but $\hat{f}$ is easily invertible since on value $\hat{y}$ the inverter can always return $\hat{y}$ itself as a possible pre-image. Still, we show below that such an $\hat{f}$ (which is only statistically correct) is a *distributionally* one-way function. We will later show how to turn a distributionally one-way function in NC^0 into a OWF in NC^0.

Definition 4.2 (Distributionally one-way function [90]**)** A polynomial-time computable function $f : \{0, 1\}^* \to \{0, 1\}^*$ is called *distributionally one-way* if there exists a positive polynomial $p(\cdot)$ such that for every (non-uniform) polynomial-time algorithm, A, and all sufficiently large n's, $\|(A(1^n, f(U_n)), f(U_n)) - (U_n, f(U_n))\| > \frac{1}{p(n)}$.

Before proving that a randomized encoding of a OWF is distributionally one-way, we need the following lemma.

Lemma 4.5 *Let $f, g : \{0, 1\}^* \to \{0, 1\}^*$ be two functions that differ on a negligible fraction of their domain; that is, $\Pr_{x \leftarrow U_n}[f(x) \neq g(x)]$ is negligible in n. Suppose that g is slightly hard to invert (but is not necessarily computable in polynomial time) and that f is computable in polynomial time. Then, f is distributionally one-way.*

Proof Let f_n and g_n be the restrictions of f and g to n-bit inputs, that is $f = \{f_n\}, g = \{g_n\}$, and define $\varepsilon(n) \stackrel{\text{def}}{=} \Pr_{x \leftarrow U_n}[f(x) \neq g(x)]$. Let $p(n)$ be the polynomial guaranteed by the assumption that g is slightly hard to invert. Assume, towards a contradiction, that f is not distributionally one-way. Then, there exists a polynomial-time algorithm, A, such that for infinitely many n's, $\|(A(1^n, f_n(U_n)), f_n(U_n)) - (U_n, f_n(U_n))\| \leq \frac{1}{2p(n)}$. Since $(U_n, f_n(U_n)) \equiv (x', f_n(U_n))$ where $x' \leftarrow f_n^{-1}(f_n(U_n))$, we get that for infinitely many n's $\|(A(1^n, f_n(U_n)), f_n(U_n)) - (x', f_n(U_n))\| \leq \frac{1}{2p(n)}$. It follows that for infinitely many n's

$$\Pr\left[A\left(1^n, f(U_n)\right) \in g_n^{-1}\left(f_n(U_n)\right)\right]$$
$$\geq \Pr_{x' \leftarrow f_n^{-1}(f_n(U_n))}\left[x' \in g_n^{-1}\left(f_n(U_n)\right)\right] - \frac{1}{2p(n)}. \tag{4.1}$$

We show that A inverts g with probability greater than $1 - \frac{1}{p(n)}$ and derive a contradiction. Specifically, for infinitely many n's we have:

$$\Pr\left[A\left(1^n, g_n(U_n)\right) \in g_n^{-1}\left(g_n(U_n)\right)\right]$$
$$\geq \Pr\left[A\left(1^n, f_n(U_n)\right) \in g_n^{-1}\left(f_n(U_n)\right)\right] - \varepsilon(n)$$
$$\geq \Pr_{x' \leftarrow f_n^{-1}(f_n(U_n))}\left[x' \in g_n^{-1}\left(f(U_n)\right)\right] - \frac{1}{2p(n)} - \varepsilon(n)$$
$$= \Pr_{x' \leftarrow f_n^{-1}(f_n(U_n))}\left[g_n\left(x'\right) = f_n(U_n)\right] - \frac{1}{2p(n)} - \varepsilon(n)$$
$$= \Pr_{x' \leftarrow f_n^{-1}(f_n(U_n))}\left[g_n\left(x'\right) = f_n\left(x'\right)\right] - \frac{1}{2p(n)} - \varepsilon(n)$$
$$= 1 - \varepsilon(n) - \frac{1}{2p(n)} - \varepsilon(n)$$
$$\geq 1 - \frac{1}{p(n)},$$

where the first inequality is due to the fact that f and g are ε-close, the second inequality uses (4.1), the second equality follows since $f(U_n) = f(x')$, the third equality is due to $x' \equiv U_n$, and the last inequality follows since ε is negligible. $\square$

We now use Lemma 4.5 to prove the distributional one-wayness of a statistically-correct encoding $\hat{f}$ based on the one-wayness of a related, perfectly correct, encoding g.

Lemma 4.6 *Suppose that $f : \{0, 1\}^* \to \{0, 1\}^*$ is a weak one-way function and $\hat{f}(x, r)$ is a computational randomized encoding of f. Then $\hat{f}$, viewed as a single-argument function, is distributionally one-way.*

Proof Let B and S be the decoder and the simulator of $\hat{f}$. Define the function $\hat{g}(x, r)$ in the following way: if $B(\hat{f}(x, r)) \neq f(x)$ then $\hat{g}(x, r) = \hat{f}(x, r')$ for some r' such that $B(\hat{f}(x, r')) = f(x)$ (such an r' exists by the statistical correctness); otherwise, $\hat{g}(x, r) = \hat{f}(x, r)$. Obviously, $\hat{g}$ is a perfectly correct encoding of f (as B perfectly decodes $f(x)$ from $\hat{g}(x, r)$). Moreover, by the statistical correctness of B, we have that $\hat{f}(x, \cdot)$ and $\hat{g}(x, \cdot)$ differ only on a negligible fraction of the r's. It follows that $\hat{g}$ is also a computationally-private encoding of f (because $\hat{g}(x, U_m) \overset{s}{\equiv} \hat{f}(x, U_m) \overset{c}{\equiv} S(f(x))$). Since f is slightly hard to invert, it follows from Lemma 4.4 that $\hat{g}$ is also slightly hard to invert. (Note that $\hat{g}$ might not be computable in polynomial time; however the proof of Lemma 4.4 only requires that the simulator's running time and the randomness complexity of $\hat{g}$ be polynomially bounded.) Finally, it follows from Lemma 4.5 that $\hat{f}$ is distributionally one-way as required. $\square$

4.3.2 Main Results

Based on the above, we derive the main theorem of this section:

Theorem 4.3 *If there exists a OWF in* **CREN** *then there exists a OWF in* $\mathbf{NC}_4^0$.

Proof Let f be a OWF in **CREN**. By Lemma 4.6, we can construct a distributional OWF $\hat{f}$ in $\mathbf{NC}^0$, and then apply a standard transformation (cf. [90, Lemma 1], [70, p. 96], [144]) to convert $\hat{f}$ to a OWF $\hat{f}'$ in $\mathbf{NC}^1$. This transformation consists of two steps: Impagliazzo and Luby's $\mathbf{NC}^1$ construction of weak OWF from distributional OWF [90], and Yao's $\mathbf{NC}^0$ construction of a (standard) OWF from a weak OWF [144] (see [70, Sect. 2.3]).[8] Since $\mathbf{NC}^1 \subseteq \mathbf{PREN}$ (Theorem 4.1), we can use Lemma 4.4 to encode $\hat{f}'$ by a OWF in $\mathbf{NC}^0$, in particular, by one with locality 4. $\square$

---

[8]We will later show a degree-preserving transformation from a distributional OWF to a OWF (Lemma 6.2); however, in the current context the standard transformation suffices.

Combining Lemmas 4.4, 3.5 and Corollary 4.1, we get a similar result for one-way permutations (OWPs).

Theorem 4.4 *If there exists a one-way permutation in* **PREN** *then there exists a one-way permutation in* $\mathbf{NC}_4^0$.

In particular, using Theorems 4.1 and 4.2, we conclude that a OWF (resp., OWP) in **NL**/*poly* or $\oplus\mathbf{L}$/*poly* (resp., $\oplus\mathbf{L}$/*poly*) implies a OWF (resp., OWP) in $\mathbf{NC}_4^0$.

Theorem 4.4 can be extended to trapdoor permutations (TDPs) provided that the perfect encoding satisfies the following *randomness reconstruction* property: given x and $\hat{f}(x,r)$, the randomness r can be efficiently recovered. If this is the case, then the trapdoor of f can be used to invert $\hat{f}(x,r)$ in polynomial time (but not in $\mathbf{NC}^0$). Firstly, we compute $f(x)$ from $\hat{f}(x,r)$ using the decoder; secondly, we use the trapdoor-inverter to compute x from $f(x)$; and finally, we use the randomness reconstruction algorithm to compute r from x and $\hat{f}(x,r)$. The randomness reconstruction property is satisfied by the randomized encodings described in Sect. 4.2 and is preserved under composition and concatenation. Thus, the existence of trapdoor permutations computable in $\mathbf{NC}_4^0$ follows from their existence in $\oplus\mathbf{L}$/*poly*.

More formally, a collection of permutations $\mathscr{F} = \{f_z : D_z \to D_z\}_{z \in Z}$ is referred to as a trapdoor permutation if there exist probabilistic polynomial-time algorithms (I, D, F, F^{-1}) with the following properties. Algorithm I is an index selector algorithm that on input 1^n selects an index z from Z and a corresponding trapdoor for f_z; algorithm D is a domain sampler that on input z samples an element from the domain D_z; F is a function evaluator that given an index z and x returns $f_z(x)$; and F^{-1} is a trapdoor-inverter that given an index z, a corresponding trapdoor t and $y \in D_z$ returns $f_z^{-1}(y)$. Additionally, the collection should be hard to invert, similarly to a standard collection of one-way permutations. (For a formal definition see [70, Definition 2.4.4].) By the above argument we derive the following theorem.

Theorem 4.5 *If there exists a trapdoor permutation* $\mathscr{F}$ *whose function evaluator* F *is in* $\oplus\mathbf{L}$/*poly then there exists a trapdoor permutation* $\hat{\mathscr{F}}$ *whose function evaluator* $\hat{F}$ *is in* $\mathbf{NC}_4^0$.

Remark 4.1 (**On Theorems 4.3, 4.4 and 4.5**)

1. **Constructiveness**. In Sect. 4.2, we give a constructive way of transforming a branching program representation of a function f into an $\mathbf{NC}^0$ circuit computing its encoding $\hat{f}$. It follows that Theorems 4.3, 4.4 can be made constructive in the following sense: there exists a polynomial-time *compiler* transforming a branching program representation of a OWF (resp., OWP) f into an $\mathbf{NC}^0$ representation of a corresponding OWF (resp., OWP) $\hat{f}$. A similar result holds for other cryptographic primitives considered in this chapter.
2. **Preservation of security, a finer look**. Loosely speaking, the main security loss in the reduction follows from the expansion of the input. (The simulator's running time only has a minor effect on the security, since it is added to the overall running-time of the adversary.) Thus, to achieve a level of security

similar to that achieved by applying f on n-bit inputs, one would need to apply $\hat{f}$ on $n + m(n)$ bits (the random input part of the encoding does not contribute to the security). Going through our constructions (bit-by-bit encoding of the output based on some size-$\ell(n)$ BPs, followed by the locality construction), we get $m(n) = l(n) \cdot \ell(n)^{O(1)}$, where $l(n)$ is the output length of f. If the degree of all nodes in the BPs is bounded by a constant, the complexity is $m(n) = O(l(n) \cdot \ell(n)^2)$. It is possible to further reduce the overhead of randomized encoding for specific representation models, such as balanced formulas, using constructions of randomizing polynomials from [49, 93].

3. **Generalizations**. The proofs of the above theorems carry over to OWF whose security holds against efficient *uniform* adversaries (inverters). The same is true for all cryptographic primitives considered in this work. The proofs also naturally extend to the case of *collections* of OWF and OWP (see Appendix 4.9 for discussion).

4. **Concrete assumptions**. The existence of a OWF in **SREN** (in fact, even in $\mathbf{NC}^1$) follows from the intractability of factoring and lattice problems [3]. The existence of a OWF *collection* in **SREN** follows from the intractability of the discrete logarithm problem. Thus, we get OWFs in $\mathbf{NC}_4^0$ under most standard cryptographic assumptions. In the case of OWP, we can get a collection of OWPs in $\mathbf{NC}_4^0$ based on discrete logarithm [36, 144] (see also Appendix 4.9) or RSA with a small exponent [127].[9] The latter assumption is also sufficient for the construction of TDP in $\mathbf{NC}_4^0$.

4.4 Pseudorandom Generators in $\mathbf{NC}^0$

A *pseudorandom generator* is an efficiently computable function $G : \{0, 1\}^n \to \{0, 1\}^{l(n)}$ such that: (1) G has a positive stretch, namely $l(n) > n$, where we refer to the function $l(n) - n$ as the *stretch* of the generator; and (2) any "computationally restricted procedure" D, called a *distinguisher*, has a negligible advantage in distinguishing $G(U_n)$ from $U_{l(n)}$. That is, $|\Pr[D(1^n, G(U_n)) = 1] - \Pr[D(1^n, U_{l(n)}) = 1]|$ is negligible in n.

Different notions of PRGs differ mainly in the computational bound imposed on D. In the default case of *cryptographic* PRGs, D can be any non-uniform polynomial-time algorithm (alternatively, polynomial-time algorithm). In the case of ε-*biased* generators, D can only compute a linear function of the output bits, namely the exclusive-or of some subset of the bits. Other types of PRGs, e.g. for space-bounded computation, have also been considered. The reader is referred to [68, Chap. 3] for a comprehensive and unified treatment of pseudorandomness.

---

[9]Rabin's factoring-based OWP collection [123] seems insufficient for our purposes, as it cannot be defined over the set of *all* strings of a given length. The standard modification (cf. [72, p. 767]) does not seem to be in $\oplus \mathbf{L}/poly$.

We start by considering cryptographic PRGs. We show that a *perfect* randomized encoding of such a PRG is also a PRG. We then obtain a similar result for other types of PRGs.

4.4.1 Cryptographic Generators

Definition 4.3 (Pseudorandom generator) A pseudorandom generator (PRG) is a polynomial-time computable function, $G : \{0, 1\}^n \to \{0, 1\}^{l(n)}$, satisfying the following two conditions:

- **Expansion.** $l(n) > n$, for all $n \in \mathbb{N}$.
- **Pseudorandomness.** The ensembles $\{G(U_n)\}_{n \in \mathbb{N}}$ and $\{U_{l(n)}\}_{n \in \mathbb{N}}$ are computationally indistinguishable.

Remark 4.2 (**PRGs with sublinear stretch**) An **NC**^{0} pseudorandom generator, G, that stretches its input by a single bit can be transformed into another **NC**^{0} PRG, G', with stretch $l'(n) - n = n^c$ for an arbitrary constant $c < 1$. This can be done by applying G on n^c blocks of n^{1-c} bits and concatenating the results. Since the output of any PRG is computationally-indistinguishable from the uniform distribution even by a polynomial number of samples (see [70, Theorem 3.2.6]), the block generator G' is also a PRG. This PRG gains a pseudorandom bit from every block, and therefore stretches $n^c n^{1-c} = n$ input bits to $n + n^c$ output bits. Obviously, G' has the same locality as G.

Remark 4.3 (**PRGs with linear stretch**) We can also transform an **NC**^0_d pseudorandom generator, $G : \{0, 1\}^n \to \{0, 1\}^{cn}$, with some linear stretch factor $c > 1$ into another **NC**^{0} PRG, $G'\{0, 1\}^n \to \{0, 1\}^{c'n}$, with arbitrary linear stretch factor $c' > 1$ and larger (but constant) output locality d'. This can be done by composing G with itself a constant number of times. That is, we let $G'(x) \overset{\text{def}}{=} G^{\lceil \log_c c' \rceil}(x)$ where $G^{i+1}(x) \overset{\text{def}}{=} G(G^i(x))$ and $G^0(x) \overset{\text{def}}{=} x$. Since the output of a PRG is pseudorandom even if it is invoked on a pseudorandom seed (see [70, p. 176]), the composed generator G' is also a PRG. Clearly, this PRG stretches n input bits to $c'n$ output bits and its locality is $d^{\lceil \log_c c' \rceil} = O(1)$.[10]

Remark 4.2 also applies to other types of generators considered in this section, and therefore we only use a crude classification of the stretch as being "sublinear", "linear" or "superlinear".

Lemma 4.7 *Let* $\hat{G} : \{0, 1\}^n \times \{0, 1\}^{m(n)} \to \{0, 1\}^{s(n)}$ *be a perfect randomized encoding of a PRG* $G : \{0, 1\}^n \to \{0, 1\}^{l(n)}$. *Then* $\hat{G}$, *viewed as a single-argument function, is also a PRG.*

---

[10]We can also increase the stretch factor by using the standard construction of Goldreich-Micali [70, Sect. 3.3.2]. In this case the locality of G' will be $d^{\lceil (c'-1)/(c-1) \rceil}$.

Proof Since $\hat{G}$ is stretch preserving, it is guaranteed to expand its seed. To prove the pseudorandomness of its output, we again use a reducibility argument. Assume, towards a contradiction, that there exists an efficient distinguisher $\hat{D}$ that distinguishes between U_s and $\hat{G}(U_n, U_m)$ with some non-negligible advantage ϕ; i.e., ϕ such that $\phi(n + m) > \frac{1}{q(n+m)}$ for some polynomial $q(\cdot)$ and infinitely many n's. We use $\hat{D}$ to obtain a distinguisher D between U_l and $G(U_n)$ as follows. On input $y \in \{0, 1\}^l$, run the balanced simulator of $\hat{G}$ on y, and invoke $\hat{D}$ on the resulting $\hat{y}$. If y is taken from U_l then the simulator, being balanced, outputs $\hat{y}$ that is distributed as U_s. On the other hand, if y is taken from $G(U_n)$ then, by Fact 2.5, the output of the simulator is distributed as $\hat{G}(U_n, U_m)$. Thus, the distinguisher D we get for G has the same advantage as the distinguisher $\hat{D}$ for $\hat{G}$. That is, the advantage of D is $\phi'(n) = \phi(n + m)$. Since $m(n)$ is polynomial, this advantage ϕ' is not only non-negligible in $n + m$ but also in n, in contradiction to the hypothesis. $\square$

Remark 4.4 (**The role of balance and stretch preservation**) Dropping either the balance or stretch preservation requirements, Lemma 4.7 would no longer hold. To see this consider the following two examples. Let G be a PRG, and let $\hat{G}(x, r) = G(x)$. Then, $\hat{G}$ is a perfectly correct, perfectly private, and balanced randomized encoding of G (the balanced simulator is $S(y) = y$). However, when r is sufficiently long, $\hat{G}$ does not expand its seed. On the other hand, we can define $\hat{G}(x, r) = G(x)0$, where r is a single random bit. Then, $\hat{G}$ is perfectly correct, perfectly private and stretch preserving, but its output is not pseudorandom.

Using Lemma 4.7, Theorem 4.1 and Corollary 4.1, we get:

Theorem 4.6 *If there exists a pseudorandom generator in* **PREN** *(in particular, in* $\oplus \mathbf{L}/poly$*) then there exists a pseudorandom generator in* $\mathbf{NC}_4^0$.

As in the case of OWF, an adversary that breaks the transformed generator $\hat{G}$ can break, in essentially the same time, the original generator G. Therefore, again, although the new PRG uses extra $m(n)$ random input bits, it is not more secure than the original generator applied to n bits. Moreover, we stress that the PRG $\hat{G}$ one gets from our construction has a sublinear stretch even if G has a large stretch. This follows from the fact that the length $m(n)$ of the random input is typically superlinear in the input length n. However, when G is in $\mathbf{NC}^0$, we can transform it into a PRG $\hat{G}$ in $\mathbf{NC}_4^0$ while (partially) preserving its stretch. Formally,

Theorem 4.7 *If there exists a pseudorandom generator with linear stretch in* $\mathbf{NC}^0$ *then there exists a pseudorandom generator with linear stretch in* $\mathbf{NC}_4^0$.

Proof Let G be a PRG with linear stretch in $\mathbf{NC}^0$. We can apply Theorem 4.1 to G and get, by Lemma 4.7, a PRG $\hat{G}$ in $\mathbf{NC}_4^0$. We now relate the stretch of $\hat{G}$ to the stretch of G. Let $n, \hat{n}$ be the input complexity of $G, \hat{G}$ (resp.), let $s, \hat{s}$ be the output complexity of $G, \hat{G}$ (resp.), and let $c \cdot n$ be the stretch of G, where c is a constant.

The generator $\hat{G}$ is stretch preserving, hence $\hat{s} - \hat{n} = s - n = c \cdot n$. Since G is in $\mathbf{NC}^0$, each of its output bits is computable by a constant size branching program and thus our construction adds only a constant number of random bits for each output bit of G. Therefore, the input length of $\hat{G}$ is linear in the input length of G. Hence, $\hat{s} - \hat{n} = s - n = c \cdot n = \hat{c} \cdot \hat{n}$ for some constant $\hat{c}$ and thus $\hat{G}$ has a linear stretch. $\square$

We can improve Theorem 4.6 by relying on a recent result of [83] which shows that any OWF can be transformed into a PRG via an $\mathbf{NC}^1$ reduction.[11]

Theorem 4.8 *If there exists a one-way function in* **CREN** *then there exists a pseudorandom generator in* $\mathbf{NC}_4^0$.

Proof A OWF in **CREN** can be first transformed into a OWF in $\mathbf{NC}^0$ (Theorem 4.3) and then, using [83], to a PRG in $\mathbf{NC}^1$. Combined with Theorem 4.6, this yields a PRG in $\mathbf{NC}_4^0$. $\square$

It follows that PRGs in $\mathbf{NC}_4^0$ can be constructed assuming the hardness of factoring, lattice problems, and discrete logarithm (in the latter case one gets only a collection of PRGs).

Remark 4.5 (**On unconditional $\mathbf{NC}^0$ reductions from PRG to OWF**) Our machinery can be used to obtain an $\mathbf{NC}^0$ reduction from a PRG to any OWF f, regardless of the complexity of f.[12] Moreover, this reduction only makes a *black-box* use of the underlying OWF f. The general idea is to encode the $\mathbf{NC}^1$ construction of [83] into a corresponding $\mathbf{NC}^0$ construction. Specifically, suppose $G(x) = g(x, f(q_1(x)), \ldots, f(q_m(x)))$ defines a black-box construction of a PRG G from a OWF f, where g is in **PREN** and the q_i's are in $\mathbf{NC}^0$. (The functions $g, q_1, \ldots, q_m$ are fixed by the reduction and do not depend on f.) Then, letting $\hat{g}((x, y_1, \ldots, y_m), r)$ be a perfect $\mathbf{NC}^0$ encoding of g, the function $\hat{G}(x, r) = \hat{g}((x, f(q_1(x)), \ldots, f(q_m(x))), r)$ perfectly encodes G, and hence defines a black-box $\mathbf{NC}^0$ reduction from a PRG to a OWF. The construction of [83] is of the form of $G(x)$ above (the functions $q_1, \ldots, q_m$ are simply projections there). Thus, $\hat{G}$ defines an $\mathbf{NC}^0$ reduction from a PRG to a OWF.

Comparison with Lower Bounds The results of [112] rules out the existence of a superlinear-stretch cryptographic PRG in $\mathbf{NC}_4^0$. Thus our $\mathbf{NC}_4^0$ cryptographic PRGs are not far from optimal despite their sublinear stretch. In addition, it is easy to see

---

[11]The seminal work of [86] gives a polynomial-time transformation which can be implemented in $\mathbf{NC}^1$ only in special cases, e.g., when the OWF is one-to-one or "regular", or in a nonuniform setting where an additional nonuniform advice of logarithmic length is employed by the construction. (See [82, 86] and [15, Remark 6.6].) Indeed, in older versions of this section [15], which predates [83] the following results (Theorem 4.8, Remark 4.5) were obtained only for the case of regular OWFs.

[12]Viola, in a concurrent work [138], obtains an $\mathbf{AC}^0$ reduction of this type.

that there is no PRG with degree 1 or locality 2 (since we can easily decide whether a given string is in the range of such a function). It seems likely that a cryptographic PRG with locality 3 and degree 2 can be constructed, but Theorem 4.6 leaves us one step short of this goal, in terms of both locality and degree.[13] See also Table 4.1. These gaps will be partially closed later in this work. Specifically, in Chap. 7 we construct a PRG with locality 4 and *linear* stretch whose security follows from a specific intractability assumption proposed by Alekhnovich in [5], while in Chap. 8 we construct a PRG with locality 3 and degree 2 under the assumption that it is hard to decode a random linear code. Moreover, the latter construction also enjoys an optimal *input* locality.

4.4.2 ε-Biased Generators

The proof of Lemma 4.7 uses the balanced simulator to transform a distinguisher for a PRG G into a distinguisher for its encoding $\hat{G}$. Therefore, if this transformation can be made linear, then the security reduction goes through also in the case of ε-biased generators.

Definition 4.4 (ε-biased generator) An *ε-biased generator* is a polynomial-time computable function, $G : \{0, 1\}^n \to \{0, 1\}^{l(n)}$, satisfying the following two conditions:

- **Expansion.** $l(n) > n$, for all $n \in \mathbb{N}$.
- **ε-bias.** For every linear function $L : \{0, 1\}^{l(n)} \to \{0, 1\}$ and all sufficiently large n's

$$\left| \Pr\left[L\big(G(U_n) \big) = 1 \right] - \Pr\left[L(U_{l(n)}) = 1 \right] \right| < \varepsilon(n)$$

(where a function L is *linear* if its degree over $\mathbb{F}_2$ is 1). By default, the function $\varepsilon(n)$ is required to be negligible.

Lemma 4.8 *Let G be an ε-biased generator and $\hat{G}$ a perfect randomized encoding of G. Assume that the balanced simulator S of $\hat{G}$ is linear in the sense that $S(y)$ outputs a randomized linear transformation of y (which is not necessarily a linear function of the simulator's randomness). Then, $\hat{G}$ is also an ε-biased generator.*

Proof Let $G : \{0, 1\}^n \to \{0, 1\}^{l(n)}$ and let $\hat{G} : \{0, 1\}^n \times \{0, 1\}^{m(n)} \to \{0, 1\}^{s(n)}$. Assume, towards a contradiction, that $\hat{G}$ is not ε-biased; that is, for some linear function $L : \{0, 1\}^{s(n)} \to \{0, 1\}$ and infinitely many n's, $|\Pr[L(\hat{G}(U_{n+m})) = 1]$

---

[13]Note that there exists a PRG with locality 3 if and only if there exists a PRG with degree 2. The "if" direction follows from Lemma 4.2 and Lemma 4.7, while the "only if" direction follows from Claim 3.1 and the fact that each output of an $\mathbf{NC}^0$ PRG must be balanced.

$-\Pr[L(U_s) = 1]| > \frac{1}{p(n+m)} > \frac{1}{p'(n)}$, where $m = m(n)$, $s = s(n)$, and $p(\cdot)$, $p'(\cdot)$ are polynomials. Using the balance property we get,

$$\left|\Pr\big[L\big(S(G(U_n))\big) = 1\big] - \Pr\big[L\big(S(U_l)\big) = 1\big]\right|$$

$$= \left|\Pr\big[L\big(\hat{G}(U_{n+m})\big) = 1\big] - \Pr\big[L(U_s) = 1\big]\right|$$

$$> \frac{1}{p'(n)},$$

where S is the balanced simulator of $\hat{G}$ and the probabilities are taken over the inputs as well as the randomness of S. By an averaging argument we can fix the randomness of S to some string ρ, and get $|\Pr[L(S_\rho(G(U_n))) = 1] - \Pr[L(S_\rho(U_{l(n)})) = 1]|$ $> \frac{1}{p'(n)}$, where S_ρ is the deterministic function defined by using the constant string ρ as the simulator's random input. By the linearity of the simulator, the function $S_\rho : \{0, 1\}^l \to \{0, 1\}^s$ is linear; therefore the composition of L and S_ρ is also linear, and so the last inequality implies that G is not ε-biased in contradiction to the hypothesis.   □

We now argue that the balanced simulators obtained in Sect. 4.2 are all linear in the above sense. In fact, these simulators satisfy a stronger property: for every fixed random input of the simulator, each bit of the simulator's output is determined by a single bit of its input. This simple structure is due to the fact that we encode non-boolean functions by concatenating the encodings of their output bits. We state here the stronger property as it will be needed in the next subsection.

Observation 4.1 *Let S be a simulator of a randomized encoding (of a function) that is obtained by concatenating simulators (i.e., S is defined as in the proof of Lemma 3.1). Then, fixing the randomness ρ of S, the simulator's computation has the following simple form: $S_\rho(y) = \sigma_1(y_1)\sigma_2(y_2)\cdots\sigma_l(y_l)$, where each σ_i maps y_i (i.e., the i-th bit of y) to one of two fixed strings. In particular, S computes a randomized degree-1 function of its input.*

Recall that the balanced simulator of the $\mathbf{NC}^0_4$ encoding for functions in $\oplus\mathbf{L}/poly$ (promised by Theorem 4.1) is obtained by concatenating the simulators of boolean functions in $\oplus\mathbf{L}/poly$. By Observation 4.1, this simulator is linear. Therefore, by Lemma 4.8, we can construct a sublinear-stretch ε-biased generator in $\mathbf{NC}^0_4$ from any ε-biased generator in $\oplus\mathbf{L}/poly$. In fact, one can easily obtain a nontrivial ε-biased generator even in $\mathbf{NC}^0_3$ by applying the locality construction to each of the bits of the degree-2 generator defined by $G(x, x') = (x, x', \langle x, x'\rangle)$, where $\langle \cdot, \cdot \rangle$ denotes inner product modulo 2. Again, the resulting encoding is obtained by concatenation and thus, by Observation 4.1 and Lemma 4.8, is also ε-biased. (This generator actually fools a much larger class of statistical tests; see Sect. 4.4.3 below.) Thus, we have:

Theorem 4.9 *There is a (sublinear-stretch) ε-biased generator in $\mathbf{NC}^0_3$.*

Building on a construction of Mossel et al., it is in fact possible to achieve linear stretch in $\mathbf{NC}_3^0$. Namely,

Theorem 4.10 *There is a linear-stretch ε-biased generator in $\mathbf{NC}_3^0$.*

Proof Mossel et al. present an ε-biased generator in $\mathbf{NC}^0$ with degree 2 and linear stretch ([112], Theorem 13).[14] Let G be their ε-biased generator. We can apply the locality construction (4.1) to G (using concatenation) and get, by Lemma 4.8 and Observation 4.1, an ε-biased generator $\hat{G}$ in $\mathbf{NC}_3^0$. We now relate the stretch of $\hat{G}$ to the stretch of G. Let $n, \hat{n}$ be the input complexity of $G, \hat{G}$ (resp.), let $s, \hat{s}$ be the output complexity of $G, \hat{G}$ (resp.), and let $c \cdot n$ be the stretch of G, where c is a constant. The generator $\hat{G}$ is stretch preserving, hence $\hat{s} - \hat{n} = s - n = c \cdot n$. Since G is in $\mathbf{NC}^0$, each of its output bits can be represented as a polynomial that has a constant number of monomials and thus the locality construction adds only a constant number of random bits for each output bit of G. Therefore, the input length of $\hat{G}$ is linear in the input length of G. Hence, $\hat{s} - \hat{n} = s - n = c \cdot n = \hat{c} \cdot \hat{n}$ for some constant $\hat{c}$ and thus $\hat{G}$ has a linear stretch. $\qquad\qquad\qquad\square$

Comparison with Lower Bounds It is not hard to see that there is no ε-biased generator with degree 1 or locality 2.[15] In [50] it was shown that there is no superlinear-stretch ε-biased generator in $\mathbf{NC}_3^0$. Thus, our linear-stretch $\mathbf{NC}_3^0$ generator (building on the one from [112]) is not only optimal with respect to locality and degree but is also essentially optimal with respect to stretch.

4.4.3 Generators for Space-Bounded Computation

We turn to the case of PRGs for space-bounded computation. A standard way of modeling a randomized space-bounded Turing machine is by having a random tape on which the machine can access the random bits one by one but cannot "go back" and view previous random bits (i.e., any bit that the machine wishes to remember, it must store in its limited memory). For the purpose of derandomizing such machines, it suffices to construct PRGs that fool any space-bounded distinguisher having a similar one-way access to its input. Following Babai et al. [23], we refer to such distinguishers as *space-bounded distinguishers*.

---

[14]In fact, the generator of [112, Theorem 13] is in nonuniform-$\mathbf{NC}_5^0$ (and it has a slightly super-linear stretch). However, a similar construction gives an ε-biased generator in *uniform* $\mathbf{NC}^0$ with degree 2 and linear stretch. (The locality of this generator is large but constant.) This can be done by replacing the probabilistic construction given in [112, Lemma 12] with a uniform construction of constant-degree bipartite expander with some "good" expansion properties – such a construction is given in [44, Theorem 7.1].

[15]A degree 1 generator contains more than n linear functions over n variables, which must be linearly dependent and thus biased. The non-existence of a 2-local generator follows from the fact that every nonlinear function of two input bits is biased.

Definition 4.5 (Space-bounded distinguisher [23]**)** A *space-$s(n)$ distinguisher* is a deterministic Turing machine M, and an infinite sequence of binary strings $a = (a_1, \ldots, a_n, \ldots)$ called the advice strings, where $|a_n| = 2^{O(s(n))}$. The machine has the following tapes: read-write work tapes, a read-only advice tape, and a read-only input tape on which the tested input string, y, is given. The input tape has a one-way mechanism to access the tested string; namely, at any point it may request the next bit of y. In addition, only $s(n)$ cells of the work tapes can be used. Given an n-bit input, y, the output of the distinguisher, $M^a(y)$, is the (binary) output of M where y is given on the input tape and a_n is given on the advice tape.

This class of distinguishers is a proper subset of the distinguishers that can be implemented by a space-$s(n)$ Turing machine with a two-way access to the input. Nevertheless, even log-space distinguishers are quite powerful, and many distinguishers fall into this category. In particular, this is true for the class of *linear* distinguishers considered in Sect. 4.4.2.

Definition 4.6 (PRG for space-bounded computation) We say that a polynomial-time computable function $G : \{0, 1\}^n \to \{0, 1\}^{l(n)}$ is a PRG for space $s(n)$ if $l(n) > n$ and $G(U_n)$ is indistinguishable from $U_{l(n)}$ to any space-$s(n)$ distinguisher. That is, for every space-$s(n)$ distinguisher M^a, the distinguishing advantage $|\Pr[M^a(G(U_n)) = 1] - \Pr[M^a(U_{l(n)}) = 1]|$ is negligible in n.

Several constructions of high-stretch PRGs for space-bounded computation exist in the literature (e.g., [23, 119]). In particular, a PRG for logspace computation from [23] can be computed using logarithmic space, and thus, by Theorem 4.1, admits an efficient perfect encoding in $\mathbf{NC}_4^0$. It can be shown (see proof of Theorem 4.11) that this $\mathbf{NC}_4^0$ encoding fools logspace distinguishers as well; hence, we can reduce the security of the randomized encoding to the security of the encoded generator, and get an $\mathbf{NC}_4^0$ PRG that fools logspace computation. However, as in the case of ε-biased generators, constructing such PRGs with a low stretch is much easier. In fact, the same "inner product" generator we used in Sect. 4.4.2 can do here as well.

Theorem 4.11 *There exists a (sublinear-stretch) PRG for sublinear-space computation in $\mathbf{NC}_3^0$.*

Proof Consider the inner product generator $G(x, x') = (x, x', \langle x, x' \rangle)$, where $x, x' \in \{0, 1\}^n$. It follows from the average-case hardness of the inner product function for two-party communication complexity [46] that G fools all sublinear-space distinguishers. (Indeed, a sublinear-space distinguisher implies a sublinear-communication protocol predicting the inner product of x and x'. Specifically, the party holding x runs the distinguisher until it finishes reading x, and then sends its configuration to the party holding x'.)

Applying the locality construction to G, we obtain a perfect encoding $\hat{G}$ in $\mathbf{NC}_3^0$. (In fact, we can apply the locality construction only to the last bit of G and leave the other outputs as they are.) We argue that $\hat{G}$ inherits the pseudorandomness of G. As before, we would like to argue that if $\hat{M}$ is a sublinear-space distinguisher breaking $\hat{G}$ and S is the balanced simulator of the encoding, then $\hat{M}(S(\cdot))$ is a sublinear-space distinguisher breaking G. Similarly to the proof of Lemma 4.8, the fact that $\hat{M}(S(\cdot))$ can be implemented in sublinear space will follow from the simple structure of S. However, in contrast to Lemma 4.8, here it does not suffice to require S to be linear and we need to rely on the stronger property guaranteed by Observation 4.1.[16]

We now formalize the above. As argued in Observation 4.1, fixing the randomness ρ of S, the simulator's computation can be written as

$$S_\rho(y) = \sigma_1(y_1)\sigma_2(y_2)\cdots\sigma_l(y_l),$$

where each σ_i maps a bit of y to one of two fixed strings. We can thus use S to turn a sublinear-space distinguisher $\hat{M}^a$ breaking $\hat{G}$ into a sublinear-space distinguisher $M^{a'}$ breaking G. Specifically, let the advice a' include, in addition to a, the $2l$ strings $\sigma_i(0), \sigma_i(1)$ corresponding to a "good" ρ which maintains the distinguishing advantage. (The existence of such ρ follows from an averaging argument.) The machine $M^{a'}(y)$ can now emulate the computation of $\hat{M}^a(S_\rho(y))$ using sublinear space and a one-way access to y by applying $\hat{M}^a$ in each step to the corresponding string $\sigma_i(y_i)$. $\qquad\square$

4.4.4 Pseudorandom Generators—Conclusion

We conclude this section with Table 4.1, which summarizes some of the PRGs constructed here as well as previous ones from [112] and highlights the remaining gaps. We will partially close these gaps in Chaps. 7 and 8 under specific *intractability* assumptions. (In particular, we will construct a PRG with locality 4 and *linear* stretch, as well as a PRG with output locality 3, input locality 3, and degree 2.)

4.5 Collision-Resistant Hashing in $\mathbf{NC}^0$

We start with a formal definition of collision-resistant hash-functions (CRHFs).

Definition 4.7 (Collision-resistant hashing) Let $\ell, \ell' : \mathbb{N} \to \mathbb{N}$ be such that $\ell(n) > \ell'(n)$ and let $Z \subseteq \{0, 1\}^*$. A collection of functions $\{h_z\}_{z\in Z}$ is said to be *collision-resistant* if the following holds:

---

[16]Indeed, in the current model of (non-uniform) space-bounded computation with *one-way* access to the input (and two-way access to the advice), there exist a boolean function $\hat{M}$ computable in sublinear space and a linear function S such that the composed function $\hat{M}(S(\cdot))$ is not computable in sublinear space. For instance, let $\hat{M}(y_1, \ldots, y_{2n}) = y_1 y_2 + y_3 y_4 + \cdots + y_{2n-1} y_{2n}$ and $S(x_1, \ldots, x_{2n}) = (x_1, x_{n+1}, x_2, x_{n+2}, \ldots, x_n, x_{2n})$.

Table 4.1 Summary of known pseudorandom generators. Results of Mossel et al. [112] appear in the *top part* and results of this chapter in the *bottom part*. A parameter is marked as optimal ($\checkmark$) if when fixing the other parameters it cannot be improved. A stretch entry is marked with $\times\!\!\!\checkmark$ if the stretch is sublinear and cannot be improved to be superlinear (but might be improved to be linear). The symbol * indicates a conditional result

Type	Stretch	Locality	Degree
ε-biased	superlinear	5	$2\checkmark$
ε-biased	$n^{\Omega(\sqrt{k})}$	large k	$\Omega(\sqrt{k})$
ε-biased	$\Omega(n^2)\checkmark$	$\Omega(n)$	$2\checkmark$
ε-biased	linear $\checkmark$	$3\checkmark$	$2\checkmark$
space	sublinear $\times\!\!\!\checkmark$	$3\checkmark$	$2\checkmark$
cryptographic*	sublinear $\times\!\!\!\checkmark$	4	3

1. There exists a probabilistic polynomial-time *key-generation* algorithm, G, that on input 1^n outputs an *index* $z \in Z$ (of a function h_z). The function h_z maps strings of length $\ell(n)$ to strings of length $\ell'(n)$.
2. There exists a polynomial-time *evaluation* algorithm that on input $z \in G(1^n)$, $x \in \{0, 1\}^{\ell(n)}$ computes $h_z(x)$.
3. Collisions are hard to find. Formally, a pair (x, x') is called a *collision* for a function h_z if $x \neq x'$ but $h_z(x) = h_z(x')$. The collision-resistance requirement states that every non-uniform polynomial-time algorithm A, that is given input $(z = G(1^n), 1^n)$, succeeds in finding a collision for h_z with a negligible probability in n (where the probability is taken over the coin tosses of both G and A).

Lemma 4.9 *Suppose* $\mathcal{H} = \{h_z\}_{z \in Z}$ *is collision resistant and* $\hat{\mathcal{H}} = \{\hat{h}_z\}_{z \in Z}$ *is a perfect randomized encoding of* $\mathcal{H}$. *Then* $\hat{\mathcal{H}}$ *is also collision resistant.*

Proof Since $\hat{h}_z$ is stretch preserving, it is guaranteed to shrink its input as h_z. The key generation algorithm G of $\mathcal{H}$ is used as the key generation algorithm of $\hat{\mathcal{H}}$. By the uniformity of the collection $\hat{\mathcal{H}}$, there exists an efficient evaluation algorithm for this collection. Finally, any collision $((x, r), (x', r'))$ under $\hat{h}_z$ (i.e., $(x, r) \neq (x', r')$ and $\hat{h}_z(x, r) = \hat{h}_z(x', r')$), defines a collision (x, x') under h_z. Indeed, perfect correctness ensures that $h_z(x) = h_z(x')$ and unique-randomness (see Lemma 3.5) ensures that $x \neq x'$. Thus, an efficient algorithm that finds collisions for $\hat{\mathcal{H}}$ with non-negligible probability yields a similar algorithm for $\mathcal{H}$. $\square$

By Lemma 4.9, Theorem 4.1 and Corollary 4.1, we get:

Theorem 4.12 *If there exists a CRHF* $\mathcal{H} = \{h_z\}_{z \in Z}$ *such that the function* $h'(z, x) \overset{\text{def}}{=} h_z(x)$ *is in* **PREN** *(in particular, in* $\oplus\mathbf{L}/poly$*), then there exists a CRHF* $\hat{\mathcal{H}} = \{\hat{h}_z\}_{z \in Z}$ *such that the mapping* $(z, y) \mapsto \hat{h}_z(y)$ *is in* $\mathbf{NC}_4^0$.

Using Theorem 4.12, we can construct CRHFs in NC^0 based on the intractability of factoring [51], discrete logarithm [122], or lattice problems [73, 125]. All these candidates are computable in NC^1 provided that some pre-computation is done by the key-generation algorithm. Note that the key generation algorithm of the resulting NC^0 CRHF is not in NC^0. For more details on NC^0 computation of collections of cryptographic primitives see Appendix 4.9.

4.6 Encryption in NC^0

We turn to the case of encryption. We first show that computational encoding preserves the security of a semantically secure encryption scheme both in the public-key and in the private-key setting. This result provides an NC^0 encryption algorithm but does not promise anything regarding the parallel complexity of the decryption process. This raises the question whether decryption can also be implemented in NC^0. In Sect. 4.6.2 we show that, except for some special settings (namely, private-key encryption which is secure only for a single message of bounded length or stateful private-key encryption), decryption in NC^0 is impossible regardless of the complexity of encryption. Finally, in Sect. 4.6.3 we explore the effect of randomized encoding on encryption schemes which enjoy stronger notions of security. In particular, we show that randomized encoding preserves security against chosen plaintext attacks (CPA) as well as a priori chosen ciphertext attacks (CCA1). However, randomized encoding does not preserve security against a posteriori chosen ciphertext attack (CCA2). Still, we show that the encoding of a CCA2-secure scheme enjoys a relaxed security property that suffices for most applications of CCA2-security.

4.6.1 Main Results

Suppose that $\mathscr{E} = (G, E, D)$ is a public-key encryption scheme, where G is a key generation algorithm, the encryption function $E(e, x, r)$ encrypts the message x using the key e and randomness r, and $D(d, y)$ decrypts the cipher y using the decryption key d. As usual, the functions G, E, D are polynomial-time computable, and the scheme provides correct decryption and satisfies indistinguishability of encryptions [80]. Let $\hat{E}$ be a randomized encoding of E, and let $\hat{D}(d, \hat{y}) \stackrel{\text{def}}{=} D(d, C(\hat{y}))$ be the composition of D with the decoder B of $\hat{E}$. We argue that the scheme $\hat{\mathscr{E}} \stackrel{\text{def}}{=} (G, \hat{E}, \hat{D})$ is also a public-key encryption scheme. The efficiency and correctness of $\hat{\mathscr{E}}$ are guaranteed by the uniformity of the encoding and its correctness. Using the efficient simulator of $\hat{E}$, we can reduce the security of $\hat{\mathscr{E}}$ to that of $\mathscr{E}$. Namely, given an efficient adversary $\hat{A}$ that distinguishes between encryptions of x and x' under $\hat{\mathscr{E}}$, we can break $\mathscr{E}$ by using the simulator to transform original ciphers into "new" ciphers, and then invoke $\hat{A}$. The same argument holds in the private-key setting. We now formalize this argument.

Definition 4.8 (Public-key encryption) A *secure public-key encryption scheme* (PKE) is a triple (G, E, D) of probabilistic polynomial-time algorithms satisfying the following conditions:

- **Viability.** On input 1^n the key generation algorithm, G, outputs a pair of keys (e, d). For every pair (e, d) such that $(e, d) \in G(1^n)$, and for every plaintext $x \in \{0, 1\}^*$, the algorithms E, D satisfy

$$\Pr\left[D\big(d, E(e, x)\big) \neq x\right] \leq \varepsilon(n)$$

  where $\varepsilon(n)$ is a negligible function and the probability is taken over the internal coin tosses of algorithms E and D.
- **Security.** For every polynomial $\ell(\cdot)$, and every family of plaintexts $\{x_n\}_{n \in \mathbb{N}}$ and $\{x'_n\}_{n \in \mathbb{N}}$ where $x_n, x'_n \in \{0, 1\}^{\ell(n)}$, it holds that

$$\left(e \leftarrow G_1(1^n), E\big(eG_1(1^n), x_n\big)\right) \overset{\mathrm{c}}{\equiv} \left(e \leftarrow G_1(1^n), E\big(eG_1(1^n), x'_n\big)\right), \qquad (4.2)$$

  where $G_1(1^n)$ denotes the first element in the pair $G(1^n)$.

The definition of a *private-key* encryption scheme is similar, except that the public key is omitted from the ensembles. That is, instead of (4.2) we require that $E(G_1(1^n), x_n) \overset{\mathrm{c}}{\equiv} E(G_1(1^n), x'_n)$. An extension to multiple-message security, where the indistinguishability requirement should hold for encryptions of polynomially many messages, follows naturally (see [72, Chap. 5] for formal definitions). In the public-key case, multiple-message security is implied by single-message security as defined above, whereas in the private-key case it is a strictly stronger notion. In the following we explicitly address only the (single-message) public-key case, but the treatment easily holds for the case of private-key encryption with multiple-message security.

Lemma 4.10 *Let $\mathscr{E} = (G, E, D)$ be a secure public-key encryption scheme, where $E(e, x, r)$ is viewed as a polynomial-time computable function that encrypts the message x using the key e and randomness r. Let $\hat{E}((e, x), (r, s)) = \hat{E}((e, x, r), s)$ be a computational randomized encoding of E and let $\hat{D}(d, \hat{y}) \overset{\text{def}}{=} D(d, B(\hat{y}))$ be the composition of D with the decoder B of $\hat{E}$. Then, the scheme $\hat{\mathscr{E}} \overset{\text{def}}{=} (G, \hat{E}, \hat{D})$ is also a secure public-key encryption scheme.*

Proof The uniformity of the encoding guarantees that the functions $\hat{E}$ and $\hat{D}$ can be efficiently computed. The viability of $\hat{\mathscr{E}}$ follows in a straightforward way from the correctness of the decoder B. Indeed, if (e, d) are in the support of $G(1^n)$, then for any plaintext x we have

$$\Pr_{r,s}\left[\hat{D}\big(d, \hat{E}(e, x, r, s)\big) \neq x\right] = \Pr_{r,s}\left[D\big(d, B(\hat{E}(e, x, r, s))\big) \neq x\right]$$

$$\leq \Pr_{r,s}\left[B\big(\hat{E}((e, x, r), s)\big) \neq E(e, x, r)\right]$$

$$+ \Pr_{r}\left[D\big(d, E(e, x, r)\big) \neq x\right]$$

$$\leq \varepsilon(n),$$

where $\varepsilon(\cdot)$ is negligible in n and the probabilities are also taken over the coin tosses of D; the first inequality follows from the union bound and the second from the viability of $\mathscr{E}$ and the statistical correctness of $\hat{E}$.

We move on to prove the security of the construction. Let S be the efficient computational simulator of $\hat{E}$. Then, for every polynomial $\ell(\cdot)$, and every family of plaintexts $\{x_n\}_{n \in \mathbb{N}}$ and $\{x'_n\}_{n \in \mathbb{N}}$ where $x_n, x'_n \in \{0, 1\}^{\ell(n)}$, it holds that

$$\left(e \leftarrow G_1(1^n), \hat{E}(e, x_n, r_n, s_n)\right) \overset{c}{\equiv} \left(e \leftarrow G_1(1^n), S\big(E(e, x_n, r_n)\big)\right)$$

$$\overset{c}{\equiv} \left(e \leftarrow G_1(1^n), S\big(E(e, x'_n, r_n)\big)\right)$$

$$\overset{c}{\equiv} \left(e \leftarrow G_1(1^n), \hat{E}(e, x'_n, r_n, s_n)\right),$$

where r_n and s_n are uniformly chosen random strings of an appropriate length. In the above the first and third transitions are due to the privacy of $\hat{E}$ and Fact 2.10, and the second transition is due to the security of $\mathscr{E}$ and Fact 2.8. Overall, the security of $\hat{\mathscr{E}}$ follows from the transitivity of the relation $\overset{c}{\equiv}$ (Fact 2.6). $\qquad\square$

In particular, if the scheme $\mathscr{E} = (G, E, D)$ enables errorless decryption and the encoding $\hat{E}$ is perfectly correct, then the scheme $\hat{\mathscr{E}}$ also enables errorless decryption. Additionally, the above lemma is easily extended to the case of private-key encryption with multiple-message security. Thus we get,

Theorem 4.13 *If there exists a secure public-key encryption scheme (respectively, a secure private-key encryption scheme) $\mathscr{E} = (G, E, D)$, such that E is in* **CREN** *(in particular, in* **NL**$/polyor \oplus$**L**$/poly$*), then there exists a secure public-key encryption scheme (respectively, a secure private-key encryption scheme) $\hat{\mathscr{E}} = (G, \hat{E}, \hat{D})$, such that $\hat{E}$ is in* $\mathbf{NC}^0_4$.

Specifically, one can construct an $\mathbf{NC}^0$ PKE based on either factoring [35, 77, 123], the Diffie-Hellman Assumption [64, 77] or lattice problems [4, 125]. (These schemes enable an $\mathbf{NC}^1$ encryption algorithm given a suitable representation of the key.) We will later (Remark 5.5) relax these assumptions and obtain a PKE in $\mathbf{NC}^0$ based on the combination of an arbitrary (polynomial-time computable) PKE and a one-way function in **CREN**.

4.6.2   On Decryption in $\mathbf{NC}^0$

Our construction provides an $\mathbf{NC}^0$ encryption algorithm but does not promise anything regarding the parallel complexity of the decryption process. This raises the question whether decryption can also be implemented in $\mathbf{NC}^0$.

4.6.2.1   Negative Results

We now show that, in many settings, decryption in $\mathbf{NC}^0$ is impossible regardless of the complexity of encryption. Here we consider standard *stateless* encryption schemes. We begin with the case of multiple-message security (in either the private-key or public-key setting). If a decryption algorithm $D(d, y)$ is in $\mathbf{NC}_k^0$, then an adversary that gets n encrypted messages can correctly guess the first bits of *all* the plaintexts (jointly) with at least 2^{-k} probability. To do so, the adversary simply guesses at random the k (or less) bits of the key d on which the first output bit of D depends, and then computes this first output bit (which is supposed to be the first plaintext bit) on each of the n ciphertexts using the subkey it guessed. Whenever the adversary guesses the k bits correctly, it succeeds to find the first bits of *all* n messages. When $n > k$, this violates the semantic security of the encryption scheme. Indeed, for the encryption scheme to be secure, the adversary's success probability (when the messages are chosen at random) can only be negligibly larger than 2^{-n}. (That is, an adversary cannot do much better than simply guessing these first bits.)

One may still hope that the decryption function may be implemented in $\mathbf{NC}^0$ for every *fixed* value of the key d; namely, that $D(d, \cdot) \in \mathbf{NC}_k^0$ for every fixed d. It turns out that this is impossible as well. Suppose that n-bit messages are encrypted by $m = m(n)$ bit ciphertexts, i.e., $D(d, \cdot) : \{0, 1\}^m \to \{0, 1\}^n$. By definition, for every d the boolean function $D_1(d, \cdot)$ which decrypts the first bit of the message depends on at most k bits. Therefore, no matter what d is, $D_1(d, \cdot)$ is taken from a small set of $(2m)^k$ possible functions. An adversary can therefore guess $D_1(d, \cdot)$ and decrypt the first bit of $t > (2m)^k$ ciphertexts with overall success probability of $1/(2m)^k$. Again, it can be shown that this violates semantic security (e.g., by taking the ciphertexts to be encryptions of random n-bit messages).

Finally, let us consider the case of a single-message private-key encryption. In this case, it is impossible to implement the decryption algorithm $D(d, y)$ in $\mathbf{NC}_k^0$ while supporting an arbitrary (polynomial) message length. Indeed, when the message length exceeds $k|d|^k$ (where $|d|$ is the length of the decryption key), there must be at least $k + 1$ bits of the output of D which depend on the same k bits of the key, in which case we are in the same situation as in our first attack. That is, we can guess the k bits of the key and learn the value of $k + 1$ bits of the message with success probability 2^{-k}. Again, if we consider a randomly chosen message, this violates semantic security.

4.6.2.2 Positive Results

In contrast to the above, if the scheme is restricted to a *single* message of a bounded length (even larger than the key) we can use our machinery to construct a private-key encryption scheme in which both encryption and decryption can be computed in $\mathbf{NC}^0$. This can be done by using the output of an $\mathbf{NC}^0$ PRG to mask the plaintext. Specifically, let $E(e, x) = G(e) \oplus x$ and $D(e, y) = y \oplus G(e)$, where e is a uniformly random key generated by the key generation algorithm and G is a PRG. Unfortunately, the resulting scheme is severely limited by the low stretch of our PRGs. This approach can be also used to give multiple message security, at the price of requiring the encryption and decryption algorithms to maintain a synchronized *state*. In such a stateful encryption scheme the encryption and decryption algorithms take an additional input and produce an additional output, corresponding to their state before and after the operation. The seed of the generator can be used, in this case, as the state of the scheme. In this setting, we can obtain multiple-message security by refreshing the seed of the generator in each invocation; e.g., when encrypting the current bit the encryption algorithm can randomly choose a new seed for the next session, encrypt it along with current bit, and send this encryption to the receiver (alternatively, see [72, Construction 5.3.3]). In the resulting scheme both encryption and decryption are $\mathbf{NC}^0$ functions whose inputs include the inner state of the algorithm.

4.6.3 Security Against CPA, CCA1 and CCA2 Attacks

In this section we address the possibility of applying our machinery to encryption schemes that enjoy stronger notions of security. In particular, we consider schemes that are secure against chosen plaintext attacks (CPA), a priori chosen ciphertext attacks (CCA1), and a posteriori chosen ciphertext attacks (CCA2). In all three attacks the adversary has to win the standard indistinguishability game (i.e., given a ciphertext $c = E(e, x_b)$ find out which of the two predefined plaintexts x_0, x_1 was encrypted), and so the actual difference lies in the power of the adversary. In a CPA attack the adversary can obtain encryptions of plaintexts of his choice (under the key being attacked), i.e., the adversary gets an oracle access to the encryption function. In CCA1 attack the adversary may also obtain decryptions of his choice (under the key being attacked), but he is allowed to do so only *before* the challenge is presented to him. In both cases, the security is preserved under randomized encoding. We briefly sketch the proof idea.

Let $\hat{A}$ be an adversary that breaks the encoding $\hat{\mathcal{E}}$ via a CPA attack (resp. CCA1 attack). We use $\hat{A}$ to obtain an adversary A that breaks the original scheme $\mathcal{E}$. As in the proof of Lemma 4.10, A uses the simulator to translate the challenge c, an encryption of the message x_b under $\mathcal{E}$, into a challenge $\hat{c}$, which is an encryption of the same message under $\hat{\mathcal{E}}$. Similarly, A answers the encryption queries of $\hat{A}$ (to the oracle $\hat{E}$) by directing these queries to the oracle E and applying the simulator to the result. Also, in the case of CCA1 attack, whenever $\hat{A}$ asks the decryption oracle $\hat{D}$ to decrypt some ciphertext $\hat{c}'$, the adversary A uses the decoder (of the encoding) to translate $\hat{c}'$ into a ciphertext c' of the same message under the scheme $\mathcal{E}$, and then uses the decryption oracle D to decrypt c'. This allows A to emulate the oracles $\hat{D}$ and $\hat{E}$, and thus to translate a successful CPA attack (resp. CCA1 attack) on the new scheme into a similar attack on the original scheme.

The situation is different in the case of a CCA2 attack. As in the case of a CCA1 attack, a CCA2 attacker has an oracle access to the decryption function corresponding to the decryption key in use; however, the adversary can query the oracle *even after* the challenge has been given to him, under the restriction that he cannot ask the oracle to decrypt the challenge c itself.

We start by observing that when applying a randomized encoding to a CCA2-secure encryption scheme, CCA2 security may be lost. Indeed, in the resulting encryption one can easily modify a given ciphertext challenge $\hat{c} = \hat{E}(e, x, r)$ into a ciphertext $\hat{c}' \neq \hat{c}$ which is also an encryption of the same message under the same encryption key. This can be done by applying the decoder (of the randomized encoding $\hat{E}$) and then the simulator on $\hat{c}$, that is $\hat{c}' = S(C(\hat{c}))$. Hence, one can break the encryption by simply asking the decryption oracle to decrypt $\hat{c}'$.

It is instructive to understand why the previous arguments fail to generalize to the case of CCA2 security. In the case of CCA1 attacks we transformed an adversary $\hat{A}$ that breaks the encoding $\hat{\mathcal{E}}$ into an adversary A for the original scheme in the following way: (1) we used the simulator to convert a challenge $c = E(e, x_b)$ into a challenge $\hat{c}$ which is an encryption of the same message under $\hat{\mathcal{E}}$; (2) when $\hat{A}$ asks $\hat{D}$ to decrypt a ciphertext $\hat{c}'$, the adversary A uses the decoder (of the encoding) to translate $\hat{c}'$ into a ciphertext c' of the same message under the scheme $\mathcal{E}$, and then asks the decryption oracle D to decrypt c'. However, recall that in a CCA2 attack the adversaries are not allowed to ask the oracle to decrypt the challenge itself (after the challenge is presented). So if $c' = c$ but $\hat{c}' \neq \hat{c}$, the adversary A cannot answer the (legitimate) query of $\hat{A}$.

To complement the above, we show that when applying a randomized encoding to a CCA2-secure encryption scheme not all is lost. Specifically, the resulting scheme still satisfies *Replayable CCA security (RCCA)*, a relaxed variant of CCA2 security that was suggested in [43]. Loosely speaking, RCCA security captures encryption schemes that are CCA2 secure except that they allow anyone to generate new ciphers that decrypt to the same value as a given ciphertext. More precisely, an RCCA attack is a CCA2 attack in which the adversary cannot ask the oracle to decrypt *any* cipher c' that decrypts to either x_0 or x_1 (cf. [43, Fig. 3]). This limitation prevents the problem raised in the CCA2 proof, in which a legitimate query for $\hat{D}$ translates by the decoder into an illegitimate query for D. That is, if $\hat{c}'$ does not decrypt under $\hat{\mathcal{E}}$ to either x_0 or x_1, then (by correctness) the ciphertext c' obtained by applying the decoder to $\hat{c}'$ does not decrypt to any of these messages either. Hence, randomized encoding preserves RCCA security. As argued in [43], RCCA security suffices in most applications of CCA2 security.

4.7 Other Cryptographic Primitives

The construction that was used for encryption can be adapted to other cryptographic primitives including (non-interactive) commitments, signatures, message authentication schemes (MACs), and non-interactive zero-knowledge proofs (for definitions see [70, 72]). In all these cases, we can replace the sender (i.e., the encrypting party, committing party, signer or prover, according to the case) with its randomized encoding and let the receiver (the decrypting party or verifier) use the decoding algorithm to translate the output of the new sender to an output of the original one. The security of the resulting scheme reduces to the security of the original one by using the efficient simulator and decoder. In fact, such a construction can also be generalized to the case of interactive protocols such as zero-knowledge proofs and interactive commitments. As in the case of encryption discussed above, this transformation results in an $\mathbf{NC}^0$ sender but does not promise anything regarding the parallel complexity of the receiver. (In all these cases, we show that it is impossible to implement the receiver in $\mathbf{NC}^0$.) An interesting feature of the case of commitment is that we can also improve the parallel complexity at the receiver's end (see below). The same holds for applications of commitment such as coin-flipping and ZK proofs. We now briefly sketch these constructions and their security proofs.

4.7.1 Signatures

Let $\mathscr{S} = (G, S, V)$ be a signature scheme, where G is a key-generation algorithm that generates the signing and verification keys (s, v), the signing function $S(s, \alpha, r)$ computes a signature β on the document α using the key s and randomness r, and the verification algorithm $V(v, \alpha, \beta)$ verifies that β is a valid signature on α using the verification key v. The three algorithms run in probabilistic polynomial time, and the scheme provides correct verification for legal signatures (ones that were produced by the signing function using the corresponding signing key). The scheme is secure (unforgeable) if it is infeasible to forge a signature in a chosen message attack. Namely, any(non-uniform) polynomial-time adversary that gets the verification key and an oracle access to the signing process $S(s, \cdot)$ fails to produce a valid signature β on a document α (with respect to the corresponding verification key v) for which it has not requested a signature from the oracle. (When the signing algorithm is probabilistic, the attacker does not have an access to the random coin tosses of the signing algorithm.)

Let $\hat{S}$ be a computational randomized encoding of S, and $\hat{V}(v, \alpha, \hat{\beta}) \stackrel{\text{def}}{=} V(v, \alpha, B(\hat{\beta}))$ be the composition of V with the decoder B of the encoding $\hat{S}$. We claim that the scheme $\hat{\mathscr{S}} \stackrel{\text{def}}{=} (G, \hat{S}, \hat{V})$ is also a signature scheme. The efficiency and correctness of $\hat{\mathscr{S}}$ follow from the uniformity of the encoding and its correctness. To prove the security of the new scheme we use the simulator to transform an attack on $\hat{\mathscr{S}}$ into an attack on $\mathscr{S}$. Specifically, given an adversary $\hat{A}$ that breaks $\hat{\mathscr{S}}$ with probability $\varepsilon(n)$, we can break $\mathscr{S}$ with probability $\varepsilon(n) - \text{neg}(n)$ as follows: Invoke $\hat{A}$ and emulate the oracle $\hat{S}$ using the simulator of the encoding and the signature oracle S. Given the output $(\alpha, \hat{\beta})$ of $\hat{A}$ (supposedly a new valid signature under $\hat{\mathscr{S}}$), use the decoder to translate the signature $\hat{\beta}$ into a signature β under $\mathscr{S}$ and output the pair (α, β).

To analyze the success probability of the new attack note that: (1) By Fact 2.9, the output of $\hat{A}$ when interacting with the emulated oracle is computationally indistinguishable from its output when interacting with the actual signing oracle $\hat{S}(s, \cdot)$. (2) The event that the pair $(\alpha, \hat{\beta})$ is a valid "forgery" under $\hat{\mathscr{S}}$ is efficiently detectable, and so, by (1), it happens in our emulation with probability $\varepsilon(n) - \text{neg}(n)$. (3) By definition, the decoder always translates a valid forgery $(\alpha, \hat{\beta})$ under $\hat{\mathscr{S}}$ into a valid forgery (α, β) under $\mathscr{S}$. Hence, if the scheme $\hat{\mathscr{S}}$ can be broken with non-negligible probability, then so can the scheme $\mathscr{S}$.

This argument can be extended to the private-key setting (i.e., in the case of MACs) as follows. First, observe that the indistinguishability in item (1) holds even for a distinguisher who knows the private-key. (As we encoded the function $S(\cdot, \cdot)$ who treats the key s as an additional argument.) In this case, item (2) holds as well, since forgery under $\hat{\mathscr{S}}$ is efficiently detectable given the private-key. Finally, the last item remains as is. We will later (Remark 5.5) show that signatures and MACs whose signing algorithm is in $\mathbf{NC}_4^0$ can be based on the existence of any one-way function in **CREN**, or, more concretely, on the intractability of factoring, the discrete logarithm problem, and lattice problems.[17]

Impossibility of $\mathbf{NC}^0$ Verification It is not hard to see that the verification algorithm V cannot be realized in $\mathbf{NC}^0$. Indeed, if $V \in \mathbf{NC}_k^0$ one can forge a signature on any document α by simply choosing a random string β' of an appropriate length. This attack succeeds with probability 2^{-k} since: (1) the verification algorithm checks the validity of the signature by reading at most k bits of the signature; and (2) the probability that β' agrees with the correct signature β on the bits which are read by V is at least 2^{-k}.

4.7.2 Commitments

A commitment scheme enables one party (a sender) to commit itself to a value while keeping it secret from another party (the receiver). Later, the sender can reveal the committed value to the receiver, and it is guaranteed that the revealed value is equal to the one determined at the commit stage.

---

[17]Our $\mathbf{NC}^0$ signing algorithm is probabilistic but this is unavoidable. Indeed, while a signing algorithm may generally be deterministic (see [72, p. 506]), an $\mathbf{NC}^0$ signing algorithm cannot be deterministic as in this case an adversary can efficiently learn it and use it to forge messages.

4.7.2.1 Non-interactive Commitments

We start with the simple case of a perfectly binding, non-interactive commitment. Such a scheme can be defined by a polynomial-time computable function $\text{SEND}(b, r)$ that outputs a commitment c to the bit b using the randomness r. We assume, w.l.o.g., that the scheme has a canonical decommit stage in which the sender reveals b by sending b and r to the receiver, who verifies that $\text{SEND}(b, r)$ is equal to the commitment c. The scheme should be both (computationally) hiding and (perfectly) binding. Hiding requires that $c = \text{SEND}(b, r)$ keeps b computationally secret, that is $\text{SEND}(0, U_n) \stackrel{c}{\equiv} \text{SEND}(1, U_n)$. Binding means that it is impossible for the sender to open its commitment in two different ways; that is, there are no r_0 and r_1 such that $\text{SEND}(0, r_0) = \text{SEND}(1, r_1)$.

Let $\widehat{\text{SEND}}(b, r, s)$ be a perfectly-correct computationally-private encoding of $\text{SEND}(b, r)$. Then $\widehat{\text{SEND}}$ defines a computationally-hiding perfectly-binding, non-interactive commitment. Hiding follows from the privacy of the encoding, as argued for the case of encryption in Lemma 4.10. Namely, it holds that

$$\widehat{\text{SEND}}(0, r, s) \stackrel{c}{\equiv} S\big(\text{SEND}(0, r, s)\big) \stackrel{c}{\equiv} S\big(\text{SEND}(1, r, s)\big) \stackrel{c}{\equiv} \widehat{\text{SEND}}(1, r, s)$$

where r and s are uniformly chosen strings of an appropriate length (the first and third transitions follow from the privacy of $\widehat{\text{SEND}}$ and Fact 2.10, while the second transition follows from the hiding of SEND and Fact 2.8). The binding property of $\widehat{\text{SEND}}$ follows from the perfect correctness; namely, if there exists an ambiguous pair $(r_0, s_0), (r_1, s_1)$ such that $\widehat{\text{SEND}}(0, r_0, s_0) = \widehat{\text{SEND}}(1, r_1, s_1)$, then by perfect correctness it holds that $\text{SEND}(0, r_0) = \text{SEND}(1, r_1)$ which contradicts the binding of the original scheme.[18] So when the encoding is in $\mathbf{NC}^0$ we get a commitment scheme whose sender is in $\mathbf{NC}^0$.

In fact, in contrast to the primitives described so far, here we also improve the parallel complexity at the receiver's end. Indeed, on input $\hat{c}, b, r, s$ the receiver's computation consists of computing $\widehat{\text{SEND}}(b, r, s)$ and comparing the result to $\hat{c}$. Assuming $\widehat{\text{SEND}}$ is in $\mathbf{NC}^0$, the receiver can be implemented by an $\mathbf{NC}^0$ circuit augmented with a single (unbounded fan-in) AND gate. We refer to this special type of $\mathbf{AC}^0$ circuit as an $\text{AND}_n \circ \mathbf{NC}^0$ circuit. This extension of $\mathbf{NC}^0$ is necessary as the locality of the function f that the receiver computes cannot be constant. (See the end of this subsection.)

Remark 4.6 (**Unconditional $\mathbf{NC}^0$ construction of non-interactive commitment from 1–1 OWF**) We can use our machinery to obtain an unconditional $\mathbf{NC}^0$ re-

---

[18] A modification of this scheme remains secure even if we replace SEND with a randomized encoding which is only *statistically*-correct. However, in this modification we cannot use the canonical decommitment stage. Instead, the receiver should verify the decommitment by applying the decoder B to $\hat{c}$ and comparing the result to the computation of the original sender; i.e., the receiver checks whether $B(\hat{c})$ equals to $\text{SEND}(b, r)$. A disadvantage of this alternative decommitment is that it does not enjoy the enhanced parallelism feature discussed below. Also the resulting scheme is only *statistically* binding.

duction from a non-interactive commitment scheme to any one-to-one OWF. More-over, this reduction only makes a *black-box* use of the underlying OWF f. As in the case of the $\mathbf{NC}^0$ reduction from PRGs to OWFs (Remark 4.5), the idea is to encode a non-adaptive black-box $\mathbf{NC}^1$ reduction into a corresponding $\mathbf{NC}^0$ construction. Specifically, the reduction of Blum [34] (instantiated with the Goldreich-Levin hard-core predicate [77]) has the following form: $\text{SEND}(b, (x, r)) = (f(x), r, \langle x, r \rangle \oplus b)$ where x, r are two random strings of length n. Whenever f is one-to-one OWF the resulting function SEND is a perfectly binding, non-interactive commitment (see [70, Construction 4.4.2]). Then, letting $\hat{g}((x, r, b), s)$ be a perfect $\mathbf{NC}^0$ encoding of $\langle x, r \rangle \oplus b$, the function $\widehat{\text{SEND}}(b, (x, r, s)) = (f(x), r, \hat{g}(x, r, b, s))$ perfectly en-codes SEND, and hence defines a black-box $\mathbf{NC}^0$ reduction from a non-interactive commitment scheme to a one-to-one OWF.

It follows that *non-interactive* commitments in $\mathbf{NC}^0$ are implied by the existence of a 1-1 OWF in **PREN** or by the existence of a non-interactive commitment in **PREN** (actually, perfect correctness and computational privacy suffice).(We will later show that such a scheme can be based on the intractability of factoring or discrete logarithm. See Remark 5.5.)

4.7.2.2 Interactive Commitments

While the existence of an arbitrary OWF is not known to imply non-interactive com-mitment scheme, it is possible to use OWFs to construct an *interactive* commitment scheme [114]. In particular, the PRG-based commitment scheme of [114] has the following simple form: First the receiver chooses a random string $k \in \{0, 1\}^{3n}$ and sends it to the sender, then the sender that wishes to commit to the bit b chooses a random string $r \in \{0, 1\}^n$ and sends the value of the function $\text{SEND}(b, k, r)$ to the receiver. (The exact definition of the function $\text{SEND}(b, k, r)$ is not impor-tant in our context.) To decommit the sender sends the randomness r and the bit b and the receiver accepts if $\text{SEND}(b, k, r)$ equals the message he had re-ceived in the commit phase. Computational hiding requires that for any string fam-ily $\{k_n\}$ where $k_n \in \{0, 1\}^{3n}$, choice of k it holds that $(k_n, \text{SEND}(0, k_n, U_n)) \overset{c}{\equiv} (k_n, \text{SEND}(1, k_n, U_n))$. Perfect binding requires that, except with negligible prob-ability (over the randomness of the receiver k), there are no r_0 and r_1 such that $\text{SEND}(0, k, r_0) = \text{SEND}(1, k, r_1)$.

Again, if we replace SEND by a computationally private perfectly correct en-coding $\widehat{\text{SEND}}$, we get a (two-round) interactive commitment scheme (this follows by combining the previous arguments with Fact 2.10). Moreover, as in the non-interactive case, when the encoding is in $\mathbf{NC}^0$ the receiver's computation in the decommit phase is in $\text{AND}_n \circ \mathbf{NC}^0$. Since the receiver's computation in the commit phase is also in $\mathbf{NC}^0$, we get an $\text{AND}_n \circ \mathbf{NC}^0$ receiver. (In Remark 5.5 we show that such a scheme can be based on the intractability of factoring, discrete logarithm or lattices problems.) As an immediate application, we obtain a constant-round proto-col for coin flipping over the phone [34] between an $\mathbf{NC}^0$ circuit and an $\text{AND}_n \circ \mathbf{NC}^0$ circuit.

4.7.2.3 Statistically Hiding Commitments

One can apply a similar transformation to other variants of commitment schemes, such as unconditionally hiding (and computationally binding) interactive commitments. (Of course, to preserve the security of such schemes the privacy of the encoding will have to be *statistical* rather than computational.) Unconditionally hiding commitments require some initialization phase, which typically involves a random key sent from the receiver to the sender. We can turn such a scheme into a similar scheme between an $\mathbf{NC}^0$ sender and an $\mathrm{AND}_n \circ \mathbf{NC}^0$ receiver, provided that it conforms to the following structure: (1) the receiver initializes the scheme by *locally* computing a random key k (say, a prime modulus and powers of two group elements for schemes based on discrete logarithm) and sending it to the sender; (2) the sender responds with a single message computed by the commitment function $\mathrm{SEND}(b, k, r)$ which is in **PREN** (actually, perfect correctness and statistical privacy suffice); (3) as in the previous case, the scheme has a canonical decommit stage in which the sender reveals b by sending b and r to the receiver, who verifies that $\mathrm{SEND}(b, k, r)$ is equal to the commitment c. Statistical hiding requires that for any string family $\{k_n\}$ where $k_n \in \{0, 1\}^n$, choice of k it holds that $(k_n, \mathrm{SEND}(0, k_n, U_{m(n)})) \overset{s}{\equiv} (k_n, \mathrm{SEND}(1, k_n, U_{m(n)}))$, where $m(n)$ is the number of random coins the sender uses. Computational binding requires that, except with negligible probability (over the randomness of the receiver k), an efficient adversary cannot find r_0 and r_1 such that $\mathrm{SEND}(0, k, r_0) = \mathrm{SEND}(1, k, r_1)$.

Using the CRHF-based commitment scheme of [53, 84], one can obtain schemes of the above type based on the intractability of factoring, discrete logarithm, and lattice problems. Given such a scheme, we replace the sender's function by its randomized encoding, and get as a result a statistically hiding commitment scheme whose sender is in $\mathbf{NC}^0$. The new scheme inherits the round complexity of the original scheme and thus consists of only two rounds of interaction. (The security proof is similar to the case of perfectly binding, non-interactive commitment, only this time we use Facts 2.5, 2.3 instead of Facts 2.10, 2.8.) If the random key k cannot be computed in $\mathrm{AND}_n \circ \mathbf{NC}^0$ (as in the case of factoring and discrete logarithm based schemes), one can compute k once and for all during the generation of the receiver's circuit and hardwire the key to the receiver's circuit. (See Appendix 4.9.)

4.7.2.4 Impossibility of an $\mathbf{NC}^0$ Receiver

We show that, in any of the above settings, the receiver cannot be realized in $\mathbf{NC}^0$. Recall that in the decommit stage the sender opens his commitment to the bit b by sending a single message (b, m) to the receiver, which accepts or rejects it according to b, m and his view v of the commitment stage. (This is the case in all the aforementioned variants.) Suppose that the receiver's computation $f(b, m, v)$ is in $\mathbf{NC}^0_k$. Consider an (honest) execution of the protocol up to the decommit stage in which the sender commits to 1, and the view of the receiver is v. There are two cases: (1) there exists an ambiguous opening m_0, m_1 for which $f(0, m_0) = f(1, m_1) = \mathsf{accept}$; and

(2) there is no ambiguous opening, i.e., for all m we have $f(0, m) = $ reject and $f(1, m_1) = $ accept for some m_1. We show that we can either break the binding property (in the first case) or the hiding property (in the second case). Indeed, in the first case the sender can choose two random strings m_0', m_1' of an appropriate length. The probability that m_0' (resp., m_1') agrees with m_0 (resp., m_1) on the bits which are read by the receiver is at least 2^{-k}. Hence, with probability 2^{-2k}, we have $f(0, m_0') = f(1, m_1') = $ accept and the binding property is violated. Now consider case (2). Let r be the substring of m which is read by the receiver. Since $|r| \le k$ the receiver can efficiently find b (before the decommit stage) by going over all possible r's and checking whether there exists an r such that $f(b, r, v) = $ accept. We conclude that the locality of the receiver's computation f should be superlogarithmic.

4.7.3 Zero-Knowledge Proofs

We move on to the case of non-interactive zero-knowledge proofs (NIZK). For simplicity, we begin with the simpler case of non-interactive zero knowledge proofs (NIZK). Such proof systems are similar to standard zero-knowledge protocols except that interaction is traded for the use of a public random string σ to which both the prover and the verifier have a read-only access. More formally, a NIZK (with an efficient prover) for an **NP** relation $R(x, w)$ is a pair of probabilistic polynomial-time algorithms (P, V) that satisfies the following properties:

- (Completeness) for every $(x, w) \in R$, it holds that $\Pr[V(x, \sigma, P(x, w, \sigma)) = 1] > 1 - \text{neg}(|x|)$;
- (Soundness) for every $x \notin L_R$ (i.e., x such that $\forall w, (x, w) \notin R$) and every prover algorithm P^* we have that $\Pr[V(x, \sigma, P^*(x, \sigma)) = 1] < \text{neg}(|x|)$;
- (Zero-knowledge) there exists a probabilistic polynomial-time simulator M such that for every string sequence $\{(x_n, w_n)\}$ where $(x_n, w_n) \in LR$ it holds that

$$\left\{\left(x_n, \sigma, P(x_n, w_n, \sigma)\right)\right\} \stackrel{c}{\equiv} \left\{M(x_n)\right\}$$

(where in all the above σ is uniformly distributed over $\{0, 1\}^{\text{poly}(|x|)}$).

Similarly to the previous cases, we can compile the prover into its computational randomized encoding $\hat{P}$, while the new verifier $\hat{V}$ uses the decoder B to translate the prover's encoded message $\hat{y}$ to the corresponding message of the original prover, and then invokes the original verifier (i.e., $\hat{V} = V(x, \sigma, B(\hat{y}))$). The completeness and soundness of the new protocol follow from the correctness of the encoding. The zero-knowledge property follows from the privacy of the encoding. That is, to simulate the new prover we define a simulator $\hat{M}$ that invokes the simulator M of the original scheme and then applies the simulator S of the encoding to the third entry of M's output. By Fact 2.10 and the privacy of $\hat{P}$ it holds that $(x, \sigma, \hat{P}(x, w, \sigma, r)) \stackrel{c}{\equiv} (x, \sigma, S(P(x, w, \sigma)))$ (where r is the randomness of the encoding $\hat{P}$) while Fact 2.8 ensures that $(x, \sigma, S(P(x, w, \sigma))) \stackrel{c}{\equiv} \hat{M}(x)$.

The above construction generalizes to *interactive* ZK-proofs with an efficient prover. In this case, we can encode the prover's computation (viewed as a function of its input, the **NP**-witness he holds, his private randomness and all the messages he has received so far), while the new receiver uses the decoder to translate the messages and then invokes the original protocol. The resulting protocol is still computational ZK proof. (The proof is similar to the case of NIZK above, but relies on Fact 2.9 instead of Fact 2.8.) The same construction works for ZK arguments (in which the soundness holds only against a computationally bounded cheating prover). When the encoding is *statistically*-private the above transformation also preserves the statistical ZK property. That is, if the original protocol provides a statistically-close simulation then so does the new protocol.

As before, this general approach does not parallelize the verifier; in fact, the verifier is now required to "work harder" and decode the prover's messages. However, we can improve the verifier's complexity by relying on specific, commitment-based, zero-knowledge protocols from the literature. For instance, in the constant-round protocol for Graph 3-Colorability of [75], the computations of the prover and the verifier consist of invoking two commitments (of both types, perfectly binding as well as statistically hiding), in addition to some $\mathbf{AC}^0$ computations. Hence, we can use the parallel commitment schemes described before to construct a constant-round protocol for 3-Colorability between an $\mathbf{AC}^0$ prover and an $\mathbf{AC}^0$ verifier. Since 3-Colorability is **NP** complete under $\mathbf{AC}^0$-reductions, we get constant-round zero-knowledge proofs in $\mathbf{AC}^0$ for every language in **NP**. (We will later show that the existence of such a protocol is implied by the intractability of factoring, the discrete logarithm problem, and lattice problems. See Remark 5.5.)

Impossibility of an $\mathbf{NC}^0$ Verifier   Again, it is impossible to obtain an $\mathbf{NC}^0$ verifier (for non-trivial languages). Suppose that we have a ZK-proof for a language L whose verifier V is in $\mathbf{NC}^0_k$. First, observe that in such a proof system the soundness error is either 0, or at least $2^{-k} = \Omega(1)$. Hence, since we require a negligible error probability, the soundness must be perfect. Thus, we can efficiently decide whether a string x is in L by letting the honest verifier interact with the simulator. More precisely, given x we use the simulator to sample a transcript of the protocol (with respect to an honest verifier), and accept x if the verifier accepts the transcript. If $x \in L$ then, except with negligible probability, the verifier should accept (as otherwise the interaction with the simulator is distinguishable from the interaction with the real prover). On the other hand, if $x \notin L$ then, due to the perfect soundness, the verifier always rejects. Therefore, $L \in \mathbf{BPP}$.

In fact, if the round complexity of the protocol is $t = O(1)$ as in the aforementioned constructions (and the verifier is in $\mathbf{NC}^0_k$), then L must be in $\mathbf{NC}^0_{tk}$. Furthermore, this is true for any interactive proof protocol, not necessarily ZK. To see this, note that the verifier computes its output based on at most tk bits of x. If $L \notin \mathbf{NC}^0_{tk}$ then there exist an input $x \in L$ and an input $y \notin L$ which agree on the tk bits read by V. Let v be a view of V which makes him accept x. Then, the same view makes V accept y, in contradiction to the perfect soundness.

4.7.4 Instance Hiding Schemes

An instance hiding scheme (IHS) allows a powerful machine (an oracle) to help a more limited user compute some function f on the user's input x; the user wishes to keep his input private and so he cannot just send it to the machine. We assume that the user is an algorithm from a low complexity class WEAK whereas the oracle is from a higher complexity class STRONG. In a (non-adaptive, single-oracle) IHS the user first transforms his input x into a (randomized) encrypted instance $y = E(x, r)$ and then asks the oracle to compute $z = g(y)$. The user should be able to recover the value of $f(x)$ from z by applying a decryption algorithm $D(x, r, z)$ (where $D \in$ WEAK) such that $D(x, r, g(E(x, r))) = f(x)$. The hiding of the scheme requires that $E(x, r)$ keeps x secret, i.e., for every string families $\{x_n\}$ and $\{x'_n\}$ (where $|x_n| = |x'_n|$), the ensembles $E(x_n, r)$ and $E(x'_n, r)$ are indistinguishable with respect to functions in STRONG. The default setting of instance hiding considered in the literature refers to a probabilistic polynomial-time user and a computationally unbounded machine. (See [60] for a survey on IHS schemes.) We will scale this down and let the user be an $\mathbf{NC}^0$ function and the oracle be a probabilistic polynomial-time machine.

The notion of randomized encoding naturally gives rise to IHS in the following way: Given f we define a related function $h(x, r) = f(x) \oplus r$ (where $|r| = |f(x)|$). Let $\hat{h}((x, r), s)$ be a computational randomized encoding of h whose decoder is B. Then, we define $E(x, (r, s)) = \hat{h}((x, r), s)$, $g(y) = B(y)$ and $D(x, r, z) = r \oplus z$. The correctness of the scheme follows from the correctness of the encoding. To prove the privacy note that, by Fact 2.10, it holds that

$$\hat{h}(x_n, r, s) \overset{\mathrm{c}}{\equiv} S\big(f(x_n) \oplus r\big) \equiv S\big(f(y_n) \oplus r\big) \overset{\mathrm{c}}{\equiv} \hat{h}(y_n, r, s).$$

Hence, we can construct such a scheme where WEAK $= \mathbf{NC}^0$ and STRONG $=$ **CREN**. (Recall that **CREN** $\subseteq$ **BPP** and thus computational privacy indeed fools a **CREN** oracle.)

4.8 Summary and Discussion

Table 4.2 summarizes the properties of randomized encoding that suffice for encoding different cryptographic primitives. (In the case of trapdoor permutations, efficient randomness recovery is also needed.) As mentioned before, in some cases it suffices to use a *computationally-private* randomized encoding. This relaxation allows us to construct (some) primitives in $\mathbf{NC}^0$ under more general assumptions. (See Theorem 5.3.)

Table 4.2 Sufficient properties for preserving the security of different primitives

Primitive	Encoding	Efficient simulator	Efficient decoder
One-way function	computational	required	–
One-way permutation	perfect	required	–
Trapdoor permutation	perfect	required	required
Pseudorandom generator	perfect	required	–
Collision-resistant hashing	perfect	–	–
Encryption (pub., priv.)	computational	required	required
Signatures, MAC	computational	required	required
Perfectly-binding commitment	perfectly correct comp. private	required	–
Statistically-hiding commitment	perfectly correct stat. private	required	–
Zero-knowledge proof	computational	required	required
Stat. ZK proof/arguments	statistical	required	required
Instance hiding	computational	required	required

4.8.1 The Case of PRFs

A PRF family $f_k(x)$ is a collection of efficiently-computable functions keyed by a secret key k, such that it is infeasible to distinguish between a function $f_k(\cdot)$ chosen uniformly at random from the PRF collection to a truly random function by an adversary that has an oracle access to the function. (See [70, Sect. 3.6] for a formal definition.) It is natural to ask why our machinery cannot be applied to pseudorandom functions (PRFs) (assuming there exists a PRF in **PREN**), as is implied from the impossibility results of Linial et al. [108]. Suppose that a PRF family $f_k(x) = f(k, x)$ is encoded by the function $\hat{f}(k, x, r)$. There are two natural ways to interpret $\hat{f}$ as a collection: (1) to incorporate the randomness into the key, i.e., $g_{k,r}(x) \stackrel{\text{def}}{=} \hat{f}(k, x, r)$; (2) to append the randomness to the argument of the collection, i.e., $h_k(x, r) \stackrel{\text{def}}{=} \hat{f}(k, x, r)$. To rule out the security of approach (1), it suffices to note that the mapping $\hat{f}(\cdot, r)$ is of degree one when r is fixed; thus, to distinguish $g_{k,r}$ from a truly random function, one can check whether the given function is affine (e.g., verify that $g_{k,r}(x) + g_{k,r}(y) = g_{k,r}(x + y) + g_{k,r}(0)$). The same attack applies to the function $h_k(x, r)$ obtained by the second approach, by fixing the randomness r. More generally, the privacy of a randomized encoding is guaranteed only when the randomness is secret and is freshly picked, thus our methodology works well for cryptographic primitives which employ fresh secret randomness in each invocation. PRFs do not fit into this category: while the key contains secret randomness, it is not freshly picked in each invocation.

We finally note that by combining the positive results regarding the existence of various primitives in $\mathbf{NC}^0$ with the fact that PRFs cannot be implemented in $\mathbf{NC}^0$, or even $\mathbf{AC}^0$ (see footnote 2), one can derive a separation between PRFs and other primitives such as PRGs. In particular, there is no $\mathbf{AC}^0$ construction of PRFs from PRGs, unless factoring is easy on the average (more generally, unless there is no OWF in **CREN**).

4.8.2 Open Problems

The results described in this chapter provide strong evidence for the possibility of cryptography in $\mathbf{NC}^0$. They are also close to optimal in terms of the exact locality that can be achieved. Still, several questions are left for further study. In particular:

- What are the minimal assumptions required for cryptography in $\mathbf{NC}^0$? For instance, does the existence of an arbitrary OWF imply the existence of OWF in $\mathbf{NC}^0$? We show that a OWF in $\mathbf{NL}/poly$ implies a OWF in $\mathbf{NC}^0$.
- Can the existence of a OWF (or PRG) in $\mathbf{NC}_3^0$ be based on general assumptions such as the ones that suffice for implementations in $\mathbf{NC}_4^0$? In Chaps. 6 and 8 we construct such a OWF (and even a PRG) under concrete intractability assumptions (e.g., the intractability of decoding a random linear code).
- Can our paradigm for achieving better parallelism be of any practical use?

The above questions motivate a closer study of the complexity of randomized encodings.

4.9 Appendix: On Collections of Cryptographic Primitives

In most cases, we view a cryptographic primitive (e.g., a OWF or a PRG) as a single function $f : \{0, 1\}^* \to \{0, 1\}^*$. However, it is often useful to consider more general variants of such primitives, defined by a *collection* of functions $\{f_z\}_{z \in Z}$, where $Z \subseteq \{0, 1\}^*$ and each f_z is defined over a finite domain D_z. The full specification of such a collection usually consists of a probabilistic polynomial-time key-generation algorithm that chooses an index z of a function (given a security parameter 1^n), a domain sampler algorithm that samples a random element from D_z given z, and a function evaluation algorithm that computes $f_z(x)$ given z and $x \in D_z$. The primitive should be secure with respect to the distribution defined by the key-generation and the domain sampler. (See a formal definition for the case of OWF in [70, Definition 2.4.3].)

Collections of primitives arise naturally in the context of parallel cryptography, as they allow us to shift "non-parallelizable" operations such as prime number selection and modular exponentiations to the key-generation stage (cf. [117]). They also fit naturally into the setting of P-uniform circuits, since the key-generation algorithm can be embedded in the algorithm generating the circuit. Thus, it will be convenient to assume that z is a description of a circuit computing f_z. When referring to a collection of functions from a given complexity class (e.g., $\mathbf{NC}^1$, $\mathbf{NC}_4^0$, or **PREN**, cf. Definition 3.7) we assume that the key generation algorithm outputs a description of a circuit from this class. In fact, one can view collections in our context as a natural relaxation of uniformity, allowing the circuit generator to be randomized. (The above discussion also applies to other P-uniform representation models we use, such as branching programs.)

Our usage of collections differs from the standard one in that we insist on D_z being the set of *all* strings of a given length (i.e., the set of all possible inputs for the circuit z) and restrict the domain sampler to be a trivial one which outputs a uniformly random string of the appropriate length. This convention guarantees that the primitive can indeed be invoked with the specified parallel complexity, and does not implicitly rely on a (possibly less parallel) domain sampler.[19] In most cases, it is possible to modify standard collections of primitives to conform to the above convention. We illustrate this by outlining a construction of an $\mathbf{NC}^1$ collection of one-way permutations based on the intractability of discrete logarithm. The key-generator, on input 1^n, samples a random prime p such that $2^{n-1} \le p < 2^n$ along with a generator g of Z_p^*, and lets z be a description of an $\mathbf{NC}^1$ circuit computing the function $f_{p,g}$ defined as follows. On an n-bit input x (viewed as an integer such that $0 \le x < 2^n$) define $f_{p,g}(x) = g^x \mod p$ if $1 \le x < p$ and $f_{p,g}(x) = x$ otherwise. It is easy to verify that $f_{p,g}$ indeed defines a permutation on $\{0,1\}^n$. Moreover, it can be computed by an $\mathbf{NC}^1$ circuit by incorporating $p, g, g^2, g^4, \ldots, g^{2^n}$ into the circuit. Finally, assuming the intractability of discrete logarithm, the above collection is *weakly* one way. It can be augmented into a collection of (strongly) one-way permutations by using the standard reduction of strong OWF to weak OWF (i.e., using $f'_{p,g}(x_1, \ldots, x_n) = (f_{p,g}(x_1), \ldots, f_{p,g}(x_n))$).

When defining the cryptographic security of a collection of primitives, it is assumed that the adversary (e.g., inverter or distinguisher) is given the key z, in addition to its input in the single-function variant of the primitive. Here one should make a distinction between "private-coin collections", where this is all of the information available to the adversary, and "public-coin collections" in which the adversary is additionally given the internal coin-tosses of the key-generator. (A similar distinction has been recently made in the specific context of collision-resistant hash-functions [89]; see also the discussion of "enhanced TDP" in [72, Appendix C.1].) The above example for a OWP collection is of the public-coin type. Any public-coin collection is also a private-coin collection, but the converse may not be true.

Summarizing, we consider cryptographic primitives in three different settings:

1. **Single function setting**. The circuit family $\{C_n\}_{n \in \mathbb{N}}$ that computes the primitive is constructed by a deterministic polynomial-time circuit generator that, given an input 1^n, outputs the circuit C_n. This is the default setting for most cryptographic primitives.
2. **Public-coin collection**. The circuit generator is a probabilistic polynomial-time algorithm that, on input 1^n, samples a circuit from a collection of circuits. The

---

[19]Note that unlike the key-generation algorithm, which can be applied "once and for all", the domain sampler should be invoked for each application of the primitive.

adversary gets as an input the circuit produced by the generator, along with the randomness used to generate it. The experiments defining the success probability of the adversary incorporate the randomness used by the generator, in addition to the other random variables. As in the single function setting, this generation step can be thought of as being done "once and for all", e.g., in a pre-processing stage. Public-coin collections are typically useful for primitives based on discrete logarithm assumptions, where a large prime group should be set up along with its generator and precomputed exponents of the generator.

3. **Private-coin collection**. Same as (2) except that the adversary does not know the randomness that was used by the circuit generator. This relaxation is typically useful for factoring-based constructions, where the adversary should not learn the trapdoor information associated with the public modulus (see [103, 117]).

We note that our general transformations apply to all of the above settings. In particular, given an $\mathbf{NC}^1$ primitive in any of these settings, we obtain a corresponding $\mathbf{NC}^0$ primitive in the same setting.

Chapter 5
Computationally Private Randomizing Polynomials and Their Applications

Abstract In this chapter, we study the notion of *computational* randomized encoding (cf. Definition 3.6) which relaxes the privacy property of statistical randomized encoding. We construct a computational encoding in $\mathbf{NC}_4^0$ for every *polynomial-time* computable function, assuming the existence of a one-way function (OWF) in **SREN**. (The latter assumption is implied by most standard intractability assumptions used in cryptography.) This result is obtained by combining a variant of Yao's *garbled circuit* technique with previous "information-theoretic" constructions of randomizing polynomials. We present several applications of computational randomized encoding. In particular, we relax the sufficient assumptions for parallel constructions of cryptographic primitives, obtain new parallel reductions between primitives, and simplify the design of constant-round protocols for multiparty computation.

5.1 Introduction

In Chap. 4 we showed that functions in $\oplus\mathbf{L}/poly$ (resp. **NL**) admit a perfect (resp. statistical) randomized encoding in $\mathbf{NC}_4^0$. A major question left open by these results is whether every *polynomial-time* computable function admits an encoding in $\mathbf{NC}^0$. In this chapter we consider the relaxed notion of *computational* randomized encoding. As we saw in Chap. 4, computationally private encodings are sufficient for most applications. Thus, settling the latter question for the relaxed notion may be viewed as a second-best alternative.

5.1.1 Overview of Results and Techniques

We construct a computationally private encoding in $\mathbf{NC}_4^0$ for every *polynomial-time* computable function, assuming the existence of a one-way function in **SREN** (which unconditionally contains $\oplus\mathbf{L}/poly$ and **NL**). We refer to the latter assumption as the "Easy OWF" (EOWF) assumption. We note that EOWF is a very mild assumption. In particular, it is implied by most concrete intractability assumptions commonly used in cryptography, such as those related to factoring, discrete logarithm, or lattice problems (see Remark 4.1, item 4). The $\mathbf{NC}^0$ encoding we obtain under the EOWF assumption has degree 3 and locality 4. Its size is nearly linear in the circuit size of the encoded function.

We now give a high-level overview of our construction. Recall that we wish to encode a polynomial-time computable function by an $\mathbf{NC}^0$ function. To do this we rely on a variant of Yao's *garbled circuit* technique [145]. Roughly speaking, Yao's technique allows a sender to efficiently "encrypt" a boolean circuit in a way that enables the receiver to compute the output of the circuit but "hides" any other information about the circuit's input. These properties resemble those required for randomized encoding.[1] Moreover, the garbled circuit enjoys a certain level of locality (or parallelism) in the sense that gates are encrypted independently of each other. Specifically, each encrypted gate is obtained by applying some cryptographic primitive (typically, a high-stretch PRG or an encryption scheme with special properties), on a constant number of (long) random strings and, possibly, a single input bit. However, the overall circuit might not have constant locality (it might not even be computable in $\mathbf{NC}$) as the cryptographic primitive being used in the gates might be sequential in nature. Thus, the bottleneck of the construction is the parallel-time complexity of the primitive being used.

In Chap. 4 we showed (via "information theoretic" randomized encoding) that under relatively mild assumptions many cryptographic primitives can be computed in $\mathbf{NC}^0$. Hence, we can try to plug one of these primitives into the garbled circuit construction in order to obtain an encoding with constant locality. However, a direct use of this approach would require stronger assumptions[2] than EOWF and result in an $\mathbf{NC}^0$ encoding with inferior parameters. Instead, we use the following variant of this approach.

Our construction consists of three steps. The first step is an $\mathbf{NC}^0$ implementation of *one-time symmetric encryption* using a cryptographic pseudorandom generator (PRG) as an oracle. (Such an encryption allows to encrypt a single message whose length may be polynomially larger than the key.) Crucially, our construction only requires a *minimal* PRG, i.e., one that stretches its seed only by one bit. (Such a PRG cannot be directly used to encrypt long messages. Also, it is not known whether such a minimal PRG can be transformed in parallel into a PRG with linear or superlinear stretch.) The second and main step of the construction relies on a variant of Yao's garbled circuit technique [145] to obtain an encoding in $\mathbf{NC}^0$ which uses one-time symmetric encryption as an oracle. By combining these two steps we get an encoding that can be computed by an $\mathbf{NC}^0$ circuit which uses a minimal PRG as an oracle. In the previous chapter (Theorem 4.8) we showed that, under the EOWF

---

[1]This similarity is not coincidental as both concepts were raised in the context of secure multiparty computation. Indeed, an information theoretic variant of Yao's garbled circuit technique was already used in [93] to construct low-degree randomized encoding for $\mathbf{NC}^1$ functions.

[2]Previous presentations of Yao's garbled circuit relied on primitives that seem less likely to allow an $\mathbf{NC}^0$ implementation. Specifically, [25, 115] require linear stretch PRG and [107] requires symmetric encryption that enjoys some additional properties.

assumption, a minimal PRG can be implemented in $\mathbf{NC}^0$. Hence, we can eliminate the oracle gates and obtain an $\mathbf{NC}^0$ encoding. This encoding can be converted into an $\mathbf{NC}^0$ encoding with degree 3 and locality 4 by applying another final step of "information-theoretic" encoding.

The above result gives rise to several types of cryptographic applications, discussed below.

5.1.1.1 Relaxed Assumptions for Cryptography in $\mathbf{NC}^0$

In Chap. 4 we showed that the existence of most cryptographic primitives in $\mathbf{NC}^0$ follows from their existence in higher complexity classes such as $\oplus\mathbf{L}/poly$, which is typically a very mild assumption. This result was obtained by combining the results on (information-theoretic) randomized encodings mentioned above with the fact that the security of most cryptographic primitives is inherited by their randomized encoding.

Using our construction of computationally private encodings, we can further relax the sufficient assumptions for cryptographic primitives in $\mathbf{NC}^0$. As we saw in Chap. 4, the security of most primitives is also inherited by their computationally private encoding. This is the case even for relatively "sophisticated" primitives such as public-key encryption, digital signatures, (computationally hiding) commitments, and (interactive or non-interactive) zero-knowledge proofs (see Table 4.2). Thus, given that these primitives at all exist,[3] their existence in $\mathbf{NC}^0$ follows from the EOWF assumption, namely from the existence of a OWF in complexity classes such as $\oplus\mathbf{L}/poly$. Previously (using the results of Chap. 4), the existence of each of these primitives in $\mathbf{NC}^0$ would only follow from the assumption that this particular primitive can be implemented in the above classes, a seemingly stronger assumption than EOWF.

It should be noted that we cannot obtain a similar result for some other primitives, such as one-way permutations and collision-resistant hash functions. The results for these primitives obtained in Chap. 4 rely on certain regularity properties of the encoding that are lost in the transition to computational privacy.

5.1.1.2 Parallel Reductions Between Cryptographic Primitives

In Chap. 4 we also obtained new $\mathbf{NC}^0$ reductions between cryptographic primitives. (Unlike the results discussed above, here we consider *unconditional* reductions that do not rely on unproven assumptions.) In particular, known $\mathbf{NC}^1$-reductions

---

[3]This condition is redundant in the case of signatures and commitments, whose existence follows from the existence of a PRG. We will later describe a stronger result for such primitives.

from PRG to one-way functions [83] can be encoded into $\mathbf{NC}^0$-reductions (see Remark 4.5). However, these $\mathbf{NC}^0$-reductions crucially rely on the very simple structure of the $\mathbf{NC}^1$-reductions from which they are derived. In particular, it is not possible to use the results of Chap. 4 for encoding general $\mathbf{NC}^1$-reductions (let alone polynomial-time reductions) into $\mathbf{NC}^0$-reductions.

As a surprising application of our technique, we get a general "compiler" that converts an arbitrary (polynomial-time) reduction from a primitive $\mathscr{P}$ to a OWF into an $\mathbf{NC}^0$-reduction from $\mathscr{P}$ to a OWF. This applies to all primitives $\mathscr{P}$ that are known to be equivalent to a one-way function, and whose security is inherited by their computationally-private encoding. In particular, we conclude that symmetric encryption,[4] commitment, and digital signatures are all $\mathbf{NC}^0$-reducible to one-way functions.

No parallel reductions of this type were previously known, even in $\mathbf{NC}$. The known construction of commitment from a PRG [114] requires a linear-stretch PRG (expanding n bits into $n + \Omega(n)$ bits), which is not known to be reducible *in parallel* to one-way functions (or to a minimal PRG). Other primitives, such as symmetric encryption and signatures, were not even known to be reducible in parallel to a polynomial-stretch PRG. For instance, the only previous parallel construction of symmetric encryption from a "low-level" primitive is based on the parallel PRF construction of [116]. This yields an $\mathbf{NC}^1$-reduction from symmetric encryption to *synthesizers*, a stronger primitive tha OWF (or even a PRG). Thus, we obtain better parallelism and at the same time rely on a weaker primitive. The price we pay is that we cannot generally guarantee parallel *decryption*. (See Sect. 5.3.2 for further discussion.)

An interesting feature of the new reductions is their *non-black-box* use of the underlying OWF. That is, the "code" of the $\mathbf{NC}^0$-reduction we get (implementing $\mathscr{P}$ using an oracle to a OWF) depends on the code of the OWF. This should be contrasted with most known reductions in cryptography, which make a black-box use of the underlying primitive. In particular, this is the case for the above-mentioned $\mathbf{NC}^0$-reductions based on Chap. 4. (See [126] for a thorough taxonomy of reductions in cryptography.)

5.1.1.3 Application to Secure Computation

The notion of randomizing polynomials (RP) was originally motivated by the goal of minimizing the round complexity of secure multiparty computation [27, 45, 78, 145]. The main relevant observations made in [92] were that: (1) the round complexity of most general protocols from the literature is related to the *degree* of the function being computed; and (2) if f is represented by a vector $\hat{f}$ of degree-d randomizing polynomials, then the task of securely computing f can be reduced to that

---

[4] By symmetric encryption we refer to (probabilistic) *stateless* encryption for multiple messages, where the parties do not maintain any state information other than the key. If parties are allowed to maintain synchronized states, symmetric encryption can be easily reduced in $\mathbf{NC}^0$ to a PRG.

of securely computing some *deterministic* degree-d function $\hat{f}'$ which is closely related to $\hat{f}$. This reduction from f to $\hat{f}'$ is fully *non-interactive*, in the sense that a protocol for f can be obtained by invoking a protocol for $\hat{f}$ and applying a *local* computation on its outputs (without additional interaction).

A useful corollary of our results is that under the EOWF assumption, the task of securely computing an *arbitrary* polynomial-time computable function f reduces (non-interactively) to that of securely computing a related degree-3 function $\hat{f}'$. This reduction is only *computationally* secure. Thus, even if the underlying protocol for $\hat{f}'$ is secure in an information-theoretic sense, the resulting protocol for f will only be computationally secure. (In contrast, previous constructions of randomizing polynomials maintained *information-theoretic* security, but only efficiently applied to restricted function classes such as $\oplus \mathbf{L}/poly$.) This reduction gives rise to new, conceptually simpler, constant-round protocols for general functions. For instance, a combination of our result with the classical "BGW protocol" [27] gives a simpler, and in some cases more efficient, alternative to the constant-round protocol of Beaver, Micali and Rogaway [25]. The RP-based solution requires slightly stronger assumption—log-space computable OWF rather than "standard" (poly-time computable) OWF—but can also lead to considerable efficiency improvements, as shown in [52].

5.1.2 Organization

In Sect. 5.2 we construct a computationally private encoding in **NC**^{0} for every polynomial-time computable function. Applications of this construction are discussed in Sect. 5.3. In particular, in Sect. 5.3.1 we relax the sufficient assumptions for parallel constructions of cryptographic primitives, in Sect. 5.3.2 we obtain new parallel reductions between primitives, and in Sect. 5.3.3 we simplify the design of constant-round protocols for multiparty computation.

5.2 Computational Encoding in **NC**^{0} for Efficient Functions

In this section we construct a perfectly correct computational encoding of degree 3 and locality 4 for every efficiently computable function. Our construction consists of three steps. In Sect. 5.2.1, we describe an **NC**^{0} implementation of one-time symmetric encryption using a *minimal* PRG as an oracle (i.e., a PRG that stretches its seed by just one bit). In Sect. 5.2.2 we describe the main step of the construction, in which we encode an arbitrary circuit using an **NC**^{0} circuit which uses one-time symmetric encryption as an oracle. This step is based on a variant of Yao's garbled circuit technique [145]. (The privacy proof of this construction is deferred to Sect. 5.2.4.) Combining the first two steps, we get a computational encoding in **NC**^{0} with an oracle to a minimal PRG. Finally, in Sect. 5.2.3, we derive the main result by relying on the existence of an "easy OWF", namely, a one-way function in **SREN**.

Remark 5.1 Recall that the definition of computational randomized encoding uses n both as an input length parameter and as a cryptographic "security parameter" quantifying computational privacy (see Definition 3.6). When describing our construction, it will be convenient to use a separate parameter k for the latter, where computational privacy will be guaranteed as long as $k \geq n^\varepsilon$ for some constant $\varepsilon > 0$.

5.2.1 From PRG to One-Time Encryption

An important tool in our construction is a one-time symmetric encryption; that is, a (probabilistic) private-key encryption that is semantically secure [80] for encrypting a single message. We describe an $\mathbf{NC}^0$-reduction from such an encryption to a minimal PRG, stretching its seed by a single bit. We start by defining minimal PRG and one-time symmetric encryption.

Definition 5.1 Let $G : \{0, 1\}^k \to \{0, 1\}^{\ell(k)}$ be a PRG (see Definition 4.3). We say that G is a *minimal* PRG if it stretches its input by one bit (i.e., $\ell(k) = k + 1$). When $\ell(k) = k + \Omega(k)$ we say that G is a *linear-stretch* PRG. We refer to G as a *polynomial-stretch* PRG if $\ell(k) = \Omega(k^c)$ for some constant $c > 1$.

Definition 5.2 (One-time symmetric encryption) A *one-time symmetric encryption scheme* is a pair (E, D), of probabilistic polynomial-time algorithms satisfying the following conditions:

- **Correctness.** For every k-bit key e and for every plaintext $m \in \{0, 1\}^*$, the algorithms E, D satisfy $D_e(E_e(m)) = m$ (where $E_e(m) \stackrel{\text{def}}{=} E(e, m)$ and similarly for D).
- **Security.** For every polynomial $\ell(\cdot)$, and every families of plaintexts $\{x_k\}_{k \in \mathbb{N}}$ and $\{x'_k\}_{k \in \mathbb{N}}$ where $x_k, x'_k \in \{0, 1\}^{\ell(k)}$, it holds that

$$\left\{E_{U_k}(x_k)\right\}_{k \in \mathbb{N}} \stackrel{c}{\equiv} \left\{E_{U_k}(x'_k)\right\}_{k \in \mathbb{N}}.$$

The integer k serves as the *security parameter* of the scheme. The scheme is said to be $\ell(\cdot)$-one-time symmetric encryption scheme if correctness and security hold with respect to plaintexts whose length is bounded by $\ell(k)$.

The above definition enables us to securely encrypt polynomially long messages under short keys. This is an important feature that will be used in our garbled circuit construction described in Sect. 5.2.2. In fact, it would suffice for our purposes to encrypt messages of some fixed polynomial[5] length, say $\ell(k) = k^2$. This could be easily done in $\mathbf{NC}^0$ if we had oracle access to a PRG with a corresponding stretch. Given such a PRG G, the encryption can be defined by $E_e(m) = G(e) \oplus m$ and the

---

[5]Applying the construction to circuits with a bounded fan-out, even linear length would suffice.

decryption by $D_e(c) = G(e) \oplus c$. However, we would like to base our construction on a PRG with a minimal stretch.

From the traditional "sequential" point of view, such a minimal PRG is equivalent to a PRG with an arbitrary polynomial stretch (cf. [70, Theorem 3.3.3]). In contrast, this is not known to be the case with respect to parallel reductions. It is not even known whether a linear-stretch PRG is $\mathbf{NC}$-reducible to a minimal PRG (see [138] for some relevant negative results). Thus, a minimal PRG is a more conservative assumption from the point of view of parallel cryptography. Moreover, unlike a PRG with linear stretch, a minimal PRG is $\mathbf{NC}^0$-reducible to one-way functions (see Remark 4.5).

The above discussion motivates a *direct* parallel construction of one-time symmetric encryption using a minimal PRG, i.e., a construction that does not rely on a "stronger" type of PRG as an intermediate step. We present such an $\mathbf{NC}^0$ construction below.

Construction 5.1 (From PRG to one-time symmetric encryption) *Let G be a minimal PRG that stretches its input by a single bit, let e be a k-bit key, and let m be a $(k + \ell)$-bit plaintext. Define the probabilistic encryption algorithm $E_e(m, (r_1, \ldots, r_{\ell-1})) \stackrel{\text{def}}{=} (G(e) \oplus r_1, G(r_1) \oplus r_2, \ldots, G(r_{\ell-2}) \oplus r_{\ell-1}, G(r_{\ell-1}) \oplus m)$, where $r_i \leftarrow U_{k+i}$ serve as the coin tosses of E. Given a ciphertext $(c_1, \ldots, c_{\ell-1})$, the deterministic decryption algorithm $D_e(\cdot)$ sets $r_0 = e$, computes $r_i = c_i \oplus G(r_{i-1})$ for $i = 1, \ldots, \ell$, and outputs r_ℓ.*

We prove the security of Construction 5.1 via a standard hybrid argument.

Lemma 5.1 *The scheme (E, D) described in Construction 5.1 is a one-time symmetric encryption scheme.*

Proof Construction 5.1 can be easily verified to satisfy the correctness requirement. We now prove the security of this scheme. Assume, towards a contradiction, that Construction 5.1 is not secure. It follows that there is a polynomial $\ell(\cdot)$ and two families of strings $x = \{x_k\}$ and $y = \{y_k\}$ where $|x_k| = |y_k| = k + \ell(k)$, such that the distribution ensembles $E_e(x_k)$ and $E_e(y_k)$ where $e \leftarrow U_k$, can be distinguished by a polynomial-size circuit family $\{A_k\}$ with non-negligible advantage $\varepsilon(k)$.

We use a hybrid argument to derive a contradiction. Fix some k. For a string m of length $k + \ell(k)$ we define for $0 \leq i \leq \ell(k)$ the distributions $H_i(m)$ in the following way. The distribution $H_0(m)$ is defined to be $E_{r_0}(m, (r_1, \ldots, r_{l-1}))$ where $r_i \leftarrow U_{k+i}$. For $1 \leq i \leq \ell(k)$, the distribution $H_i(m)$ is defined exactly as $H_{i-1}(m)$ only that the string $G(r_{i-1})$ is replaced with a random string w_{i-1}, which is one bit longer than r_{i-1} (that is, $w_{i-1} \leftarrow U_{k+i}$). Observe that for every $m \in \{0, 1\}^{k+\ell(k)}$, all the $\ell(k)$ strings of the hybrid $H_{\ell(k)}(m)$ are distributed uniformly and independently (each of them is the result of XOR with a fresh random string w_i). Therefore, in particular, $H_{\ell(k)}(x_k) \equiv H_{\ell(k)}(y_k)$. Since $H_0(x_k) \equiv E_e(x_k)$ as well as $H_0(y_k) \equiv E_e(y_k)$, it follows that our distinguisher A_k distinguishes, w.l.o.g., between $H_{\ell(k)}(x_k)$ and $H_0(x_k)$ with at least $\varepsilon(k)/2$ advantage. Then, since there are $\ell(k)$ hybrids, there must be $1 \le i \le \ell(k)$ such that the neighboring hybrids, $H_{i-1}(x_k)$, $H_i(x_k)$, can be distinguished by A_k with $\frac{\varepsilon(k)}{2\ell(k)}$ advantage.

We now show how to use A_k to distinguish a randomly chosen string from an output of the pseudorandom generator. Given a string z of length $k + i$ (that is either sampled from $G(U_{k+i-1})$ or from U_{k+i}), we uniformly choose the strings $r_j \in \{0, 1\}^{k+j}$ for $j = 1, \ldots, \ell(k) - 1$. We feed A_k with the sample $(r_1, \ldots, r_{i-1}, z \oplus r_i, G(r_i) \oplus r_{i+1}, \ldots, G(r_{\ell(k)-1}) \oplus x_k)$. If z is a uniformly chosen string then the above distribution is equivalent to $H_i(x_k)$. On the other hand, if z is drawn from $G(U_i)$ then the result is distributed exactly as $H_{i-1}(x_k)$, since each of the first $i - 1$ entries of $H_{i-1}(x_k)$ is distributed uniformly and independently of the remaining entries (each of these entries was XOR-ed with a fresh and unique random w_j). Hence, we constructed an adversary that breaks the PRG with non-negligible advantage $\frac{\varepsilon(k)}{2\ell(k)}$, deriving a contradiction. $\square$

Since the encryption algorithm described in Construction 5.1 is indeed an $\mathbf{NC}^0$ circuit with oracle access to a minimal PRG, we get the following lemma.

Lemma 5.2 *Let G be a minimal PRG. Then, there exists one-time symmetric encryption scheme (E, D) in which the encryption function E is in $\mathbf{NC}^0[G]$.*

Note that the decryption algorithm of the above construction is sequential. We can parallelize it (without harming the parallelization of the encryption) at the expense of strengthening the assumption we use.

Claim 5.1 *Let PG (resp. LG) be a polynomial-stretch (resp. linear-stretch) PRG. Then, for every polynomial $p(\cdot)$ there exists a $p(\cdot)$-one-time symmetric encryption scheme (E, D) such that $E \in \mathbf{NC}^0[PG]$ and $D \in \mathbf{NC}^0[PG]$ (resp. $E \in \mathbf{NC}^0[LG]$ and $D \in \mathbf{NC}^1[LG]$).*

Proof Use Construction 5.1 (where $|r_i| = |G(r_{i-1})|$). When the stretch of G is polynomial (resp. linear) the construction requires only $O(1)$ (resp. $O(\log k)$) invocations of G, and therefore, so does the decryption algorithm. $\square$

5.2.2 From One-Time Encryption to Computational Encoding

Let $f = \{f_n : \{0, 1\}^n \to \{0, 1\}^{\ell(n)}\}_{n \in \mathbb{N}}$ be a polynomial-time computable function, computed by the uniform circuit family $\{C_n\}_{n \in \mathbb{N}}$. We use a one-time symmetric encryption scheme (E, D) as a black box to encode f by a perfectly correct computational encoding $\hat{f} = \{\hat{f}_n\}_{n \in \mathbb{N}}$. Each $\hat{f}_n$ will be an $\mathbf{NC}^0$ circuit with an oracle access to the encryption algorithm E, where the latter is viewed as a function of the key, the message, and its random coin tosses. The construction uses a variant of Yao's garbled circuit technique [145]. Our notation and terminology for this section borrow from previous presentations of Yao's construction in [107, 115, 128].[6] Before we describe the actual encoding it will be convenient to think of the following "physical" analog that uses locks and boxes.

A Physical Encoding   To each wire of the circuit we assign a pair of keys: a 0-key that represents the value 0 and a 1-key that represents the value 1. For each of these pairs we randomly color one key black and the other key white. This way, given a key one cannot tell which bit it represents (since the coloring is random). For every gate of the circuit, the encoding consists of four double-locked boxes—a white-white box (which is locked by the white keys of the wires that enter the gate), a white-black box (locked by the white key of the left incoming wire and the black key of the right incoming wire), a black-white box (locked by the black key of the left incoming wire and the white key of the right incoming wire) and a black-black box (locked by the black keys of the incoming wires). Inside each box we put one of the keys of the gate's output wires. Specifically, if a box is locked by the keys that represent the values α, β then for every outgoing wire we put in the box the key that represents the bit $g(\alpha, \beta)$, where g is the function that the gate computes. For example, if the gate is an OR gate then the box which is locked by the incoming keys that represent the bits $(0, 1)$ contains all the 1-keys of the outgoing wires. So if one has a single key for each of the incoming wires, he can open only one box and get a single key for each of the outgoing wires. Moreover, as noted before, holding these keys does not reveal any information about the bits they represent.

Now, fix some input x for f_n. For each wire, exactly one of the keys corresponds to the value of the wire (induced by x); we refer to this key as the *active key* and to the second key as the *inactive* key. We include in the encoding of $f_n(x)$ the active keys of the input wires. (This is the only place in which the encoding depends on the input x.) Using these keys and the locked boxes as described above, one can obtain the active keys of all the wires by opening the corresponding boxes in a bottom-to-top order. To make this information useful (i.e., to enable decoding of $f_n(x)$), we append to the encoding the semantics of the output wires; namely, for each output wire we expose whether the 1-key is white or black. Hence, the knowledge of the active key of an output wire reveals the value of the wire.

The Actual Encoding   The actual encoding is analogous to the above physical encoding. We let random strings play the role of physical keys. Instead of locking a value in a double-locked box, we encrypt it under the XOR of two keys. Before formally defining the construction, we need the following notation. Denote by $x = (x_1, \ldots, x_n)$ the input for f_n. Let $k = k(n)$ be a security parameter which may be set to n^ε for an arbitrary positive constant ε (see Remark 5.1). Let $\Gamma(n)$ denote the number of gates in C_n. For every $1 \leq i \leq |C_n|$, denote by $b_i(x)$ the value of the i-th

---

[6]Security proofs for variants of this construction were given implicitly in [107, 128, 134] in the context of secure computation. However, they cannot be directly used in our context for different reasons. In particular, the analysis of [107] relies on a special form of symmetric encryption and does not achieve perfect correctness, while that of [128, 134] relies on a linear-stretch PRG.

wire induced by the input x; when x is clear from the context we simply use b_i to denote the wire's value.

Our encoding $\hat{f}_n(x, (r, W))$ consists of random inputs of two types: $|C_n|$ pairs of strings $W_i^0, W_i^1 \in \{0, 1\}^{2k}$, and $|C_n|$ bits (referred to as masks) denoted $r_1, \ldots, r_{|C_n|}$.[7] The strings W_i^0, W_i^1 will serve as the 0-key and the 1-key of the i-th wire, while the bit r_i will determine which of these keys is the black key. We use c_i to denote the value of wire i masked by r_i; namely, $c_i = b_i \oplus r_i$. Thus, c_i is the color of the active key of the i-th wire (with respect to the input x). As before, the encoding $\hat{f}_n(x, (r, W))$ will reveal each active key $W_i^{b_i}$ and its color c_i but will hide the inactive keys $W_i^{1-b_i}$ and the masks r_i of all the wires (except the masks of the output wires). Intuitively, since the active keys and inactive keys are distributed identically, the knowledge of an active key $W_i^{b_i}$ does not reveal the value b_i.

The encoding $\hat{f}_n$ consists of the concatenation of $O(|C_n|)$ functions, which include several entries for each gate and for each input and output wire. In what follows $\oplus$ denotes bitwise-xor on strings; when we want to emphasize that the operation is applied to single bits we will usually denote it by either $+$ or $-$. We use $\circ$ to denote concatenation. For every $\beta \in \{0, 1\}$ and every i, we view the string W_i^β as if it is partitioned into two equal-size parts denoted $W_i^{\beta,0}, W_i^{\beta,1}$.

Construction 5.2 *Let C_n be a circuit that computes f_n. Then, we define $\hat{f}_n(x, (r, W))$ to be the concatenation of the following functions of $(x, (r, W))$.*
Input wires: *For an input wire i, labeled by a literal ℓ (either some variable x_u or its negation) we append the function $W_i^\ell \circ (\ell + r_i)$.*
Gates: *Let $t \in [\Gamma(n)]$ be a gate that computes the function $g \in \{\text{AND}, \text{OR}\}$ with input wires i, j and output wires $y_1, \ldots, y_m$. We associate with this gate 4 functions that are referred to as gate labels. Specifically, for each of the 4 choices of $a_i, a_j \in \{0, 1\}$, we define a corresponding function $Q_t^{a_i, a_j}$. This function can be thought of as the box whose color is (a_i, a_j). It is defined as follows:*

$$Q_t^{a_i, a_j}(r, W) \stackrel{\text{def}}{=} E_{W_i^{a_i - r_i, a_j} \oplus W_j^{a_j - r_j, a_i}} \left(W_{y_1}^{g(a_i - r_i, a_j - r_j)} \circ \left(g(a_i - r_i, a_j - r_j) + r_{y_1} \right) \right.$$

$$\left. \circ \cdots \circ W_{y_m}^{g(a_i - r_i, a_j - r_j)} \circ \left(g(a_i - r_i, a_j - r_j) + r_{y_m} \right) \right), \qquad (5.1)$$

where E is a one-time symmetric encryption algorithm. (For simplicity, the randomness of E is omitted.) That is, the colored keys of all the output wires of this gate are encrypted under a key that depends on the keys of the input wires of the gate. Note that $Q_t^{a_i, a_j}$ depends only on the random inputs. We refer to the label $Q_t^{c_i, c_j}$ that is indexed by the colors of the active keys of the input wires as an active label, *and to the other three labels as the* inactive labels.
Output wires: *For each output wire i of the circuit, we add the mask of this wire r_i.*

---

[7]In fact, each application of the encryption scheme will use some additional random bits. To simplify notation, we keep these random inputs implicit.

It is not hard to verify that $\hat{f}_n$ is in **NC**$^0[E]$. In particular, a term of the form W_i^ℓ is a 3-local function of W_i^0, W_i^1 and ℓ, since its j-th bit depends on the j-th bit of W_i^0, the j-th bit of W_i^1 and on the literal ℓ. Similarly, the keys that are used in the encryptions are 8-local functions, and the arguments to the encryption are 6-local functions of (r, W).

We will now analyze the complexity of $\hat{f}_n$. The output complexity and randomness complexity of $\hat{f}$ are both dominated by the complexity of the gate labels. Generally, the complexity of these functions is poly($|C_n| \cdot k$) (since the encryption E is computable in polynomial time).[8] However, when the circuit C_n has bounded fan-out (say 2) each invocation of the encryption uses poly(k) random bits and outputs poly(k) bits. Hence, the overall complexity is $O(|C_n|) \cdot \text{poly}(k) = O(|C_n| \cdot n^\varepsilon)$ for an arbitrary constant $\varepsilon > 0$. Since any circuit with unbounded fan-out of size $|C_n|$ can be (efficiently) transformed into a bounded-fanout circuit whose size is $O(|C_n|)$ (at the price of a logarithmic factor in the depth), we get an encoding of size $O(|C_n| \cdot n^\varepsilon)$ for every (unbounded fan-out) circuit family $\{C_n\}$.

Let $\mu(n), s(n)$ be the randomness complexity and the output complexity of $\hat{f}_n$ respectively. We claim that the function family $\hat{f} = \{\hat{f}_n : \{0, 1\}^n \times \{0, 1\}^{\mu(n)} \to \{0, 1\}^{s(n)}\}_{n \in \mathbb{N}}$ defined above is indeed a computationally randomized encoding of the family f. We start with perfect correctness.

Lemma 5.3 (Perfect correctness) *There exists a polynomial-time decoder algorithm B such that for every $n \in \mathbb{N}$ and every $x \in \{0, 1\}^n$ and $(r, W) \in \{0, 1\}^{\mu(n)}$, it holds that*

$$B\big(1^n, \hat{f}_n(x, (r, W))\big) = f_n(x).$$

Proof Let $\alpha = \hat{f}_n(x, (r, W))$ for some $x \in \{0, 1\}^n$ and $(r, W) \in \{0, 1\}^{\mu(n)}$. Given α, our decoder computes, for every wire i, the active key $W_i^{b_i}$ and its color c_i. Then, for an output wire i, the decoder retrieves the mask r_i from α and computes the corresponding output bit of $f_n(x)$; i.e., outputs $b_i = c_i - r_i$. (Recall that the masks of the output wires are given explicitly as part of α.) The active keys and their colors are computed by scanning the circuit from bottom to top.

For an input wire i the desired value, $W_i^{b_i} \circ c_i$, is given as part of α. Next, consider a wire y that goes out of a gate t, and assume that we have already computed the desired values of the input wires i, j of this gate. We use the colors c_i, c_j of the active keys of the input wires to select the active label $Q_t^{c_i, c_j}$ of the gate t (and ignore the other 3 inactive labels of this gate). Consider this label as in (5.1); recall that this cipher was encrypted under the key $W_i^{c_i - r_i, c_j} \oplus W_j^{c_j - r_j, c_i} = W_i^{b_i, c_j} \oplus W_j^{b_j, c_i}$. Since we have already computed the values c_i, c_j, $W_i^{b_i}$ and $W_j^{b_j}$, we can decrypt the label $Q_t^{c_i, c_j}$ (by applying the decryption algorithm D). Hence, we can recover the

---

[8]Specifically, the encryption is always invoked on messages whose length is bounded by $\ell(n) \stackrel{\text{def}}{=} O(|C_n| \cdot k)$, hence we can use $\ell(n)$-one-time symmetric encryption.

encrypted plaintext, that includes, in particular, the value $W_y^{g(b_i,b_j)} \circ (g(b_i,b_j)+r_y)$, where g is the function that gate t computes. Since by definition $b_y = g(b_i,b_j)$, the decrypted string contains the desired value. $\square$

Remark 5.2 By the description of the decoder it follows that if the circuit C_n is in $\mathbf{NC}^i$, then the decoder is in $\mathbf{NC}^i[D]$, where D is the decryption algorithm. In particular if D is in $\mathbf{NC}^j$ then the decoder is in $\mathbf{NC}^{i+j}$. This fact will be useful for some of the applications discussed in Sect. 5.3.

To argue computational privacy we need to prove the following lemma, whose proof is deferred to Sect. 5.2.4.

Lemma 5.4 (Computational privacy) *There exists a probabilistic polynomial-time simulator S, such that for any family of strings $\{x_n\}_{n\in\mathbb{N}}$, $|x_n| = n$, it holds that $S(1^n, f_n(x_n)) \stackrel{c}{\equiv} \hat{f}_n(x_n, U_{\mu(n)})$.*

*Remark 5.3 (**Information-theoretic variant**)* Construction 5.2 can be instantiated with a *perfect* (information-theoretic) encryption scheme, yielding a perfectly private randomized encoding. (The privacy proof given in Sect. 5.2.4 can be easily modified to treat this case.) However, in such an encryption the key must be as long as the encrypted message [130]. It follows that the wires' key length grows exponentially with their distance from the outputs, rendering the construction efficient only for $\mathbf{NC}^1$ circuits. This information-theoretic variant of the garbled circuit construction was previously suggested in [93]. We will use it in Sect. 5.2.3 for obtaining a computational encoding with a parallel decoder.

5.2.3 Main Results

Combining Lemmas 5.3, 5.4, and 5.2 we get an $\mathbf{NC}^0$ encoding of any efficiently computable function using an oracle to a minimal PRG. Since the latter primitive is $\mathbf{NC}^0$-reducible to a OWF (Remark 4.5) we derive the following theorem.

Theorem 5.1 *Suppose f is computed by a uniform family $\{C_n\}$ of polynomial-size circuits. Let H be a OWF. Then, f admits a perfectly correct computational encoding $\hat{f}$ in $\mathbf{NC}^0[H]$. The complexity of $\hat{f}$ is $O(|C_n| \cdot n^\varepsilon)$ (for an arbitrary constant $\varepsilon > 0$).*

We turn to the question of eliminating the OWF oracles. We follow the natural approach of replacing each oracle with an $\mathbf{NC}^0$ implementation. Using Theorem 4.3, a OWF in $\mathbf{NC}^0$ is implied by a OWF in $\mathbf{SREN}$ or even $\mathbf{CREN}$. Thus, we can base our main theorem on the following "easy OWF" assumption.

Assumption 5.1 (Easy OWF (EOWF)) *There exists a OWF in $\mathbf{CREN}$.*

EOWF is a very mild assumption. It is implied by the existence of OWFs in $\mathbf{NC}^1$, $\oplus\mathbf{L}/poly$ or $\mathbf{NL}$ since these classes are unconditionally contained in $\mathbf{SREN}$ (Theorems 4.1, 4.2). It is also implied by most concrete cryptographic intractability assumptions (see Remark 4.1, item 4).

Combining Theorem 5.1 with the EOWF assumption, we get a computational encoding in $\mathbf{NC}^0$ for every efficiently computable function. To optimize its parameters we apply a final step of perfect encoding, yielding a computational encoding with degree 3 and locality 4 (see Corollary 4.1). Thus, we get the following main theorem.

Theorem 5.2 *Suppose f is computed by a uniform family $\{C_n\}$ of polynomial-size circuits. Then, under the EOWF assumption, f admits a perfectly correct computational encoding $\hat{f}$ of degree 3, locality 4 and complexity $O(|C_n| \cdot n^\varepsilon)$ (for an arbitrary constant $\varepsilon > 0$).*

Corollary 5.1 *Under the EOWF assumption, $\mathbf{CREN} = \mathbf{BPP}$.*

Proof Let $f(x)$ be a function in $\mathbf{BPP}$. It follows that there exists a function $f'(x, z) \in \mathbf{P}$ such that for every $x \in \{0, 1\}^n$ it holds that $\Pr_z[f'(x, z) \neq f(x)] \leq 2^{-n}$. Let $\hat{f}'((x, z), r)$ be the $\mathbf{NC}^0$ computational encoding of f' promised by Theorem 5.2. Since f' is a statistical encoding of f (the simulator and the decoder are simply the identity functions), it follows from Lemma 3.4 that $\hat{f}(x, (z, r)) \stackrel{\text{def}}{=} \hat{f}'((x, z), r)$ is a computational encoding of f in $\mathbf{NC}^0$.

Conversely, suppose $f \in \mathbf{CREN}$ and let $\hat{f}$ be an $\mathbf{NC}^0$ computational encoding of f. A BPP algorithm for f can be obtained by first computing $\hat{y} = \hat{f}(x, r)$ on a random r and then invoking the decoder on $\hat{y}$ to obtain the output $y = f(x)$ with high probability. $\square$

On the Parallel Complexity of the Decoder As we shall see in Sect. 5.3, it is sometimes useful to obtain a computational encoding whose decoder is also parallelized. Recall that if the circuit computing f is an $\mathbf{NC}^i$ circuit and the decryption algorithm (used in the construction) is in $\mathbf{NC}^j$, we obtain a parallel decoder in $\mathbf{NC}^{i+j}$ (see Remark 5.2). Unfortunately, we cannot use the parallel symmetric encryption scheme of Construction 5.1 for this purpose because of its sequential decryption.

We can get around this problem by strengthening the EOWF assumption. Suppose we have a *polynomial-stretch* PRG in $\mathbf{NC}^1$. (This is implied by some standard cryptographic assumptions, see [117].) In such a case, by Claim 5.1, we can obtain a one-time symmetric encryption scheme (E, D) (for messages of a fixed polynomial length) in which both E and D are in $\mathbf{NC}^1$. Our goal is to turn this into a scheme $(\hat{E}, \hat{D})$ in which the encryption $\hat{E}$ is in $\mathbf{NC}^0$ and the decryption is still in $\mathbf{NC}^1$. We achieve this by applying to (E, D) the encoding given by the information-theoretic variant of the garbled circuit construction (see Remark 5.3 or [93]). That is, $\hat{E}$ is a (perfectly correct and private) $\mathbf{NC}^0$ encoding of E, and $\hat{D}$ is obtained by composing D with the decoder of the information-theoretic garbled circuit. (The resulting scheme $(\hat{E}, \hat{D})$ is still a secure encryption scheme, see Theorem 4.13.) Since the symmetric encryption (E', D') employed by the information-theoretic garbled circuit is in $\mathbf{NC}^0$, its decoder can be implemented in $\mathbf{NC}^1[D'] = \mathbf{NC}^1$. Thus, $\hat{D}$ is also in $\mathbf{NC}^1$ (as $\mathbf{NC}^0[\text{decoder}] = \mathbf{NC}^1$). Combining this encryption scheme with Construction 5.2, we get a computational encoding of a function $f \in \mathbf{NC}^i$ with encoding in $\mathbf{NC}^0$ and decoding in $\mathbf{NC}^{i+1}$. Assuming there exists a linear-stretch PRG in $\mathbf{NC}^1$, we can use a similar argument to obtain an $\mathbf{NC}^0$ encoding for f whose decoding in $\mathbf{NC}^{i+2}$. (In this case we use the linear-PRG part of Claim 5.1.) Summarizing, we have the following.

Claim 5.2 *Suppose there exists a PRG with polynomial stretch (resp. linear stretch) in $\mathbf{NC}^1$. Then, every function $f \in \mathbf{NC}^i$ admits a perfectly-correct computational encoding in $\mathbf{NC}^0$ whose decoder is in $\mathbf{NC}^{i+1}$ (resp. $\mathbf{NC}^{i+2}$).*

5.2.4 Proof of Lemma 5.4

5.2.4.1 The Simulator

We start with the description of the simulator S. Given 1^n and $f_n(x)$, for some $x \in \{0, 1\}^n$, the simulator chooses, for every wire i of the circuit C_n, an active key and a color; namely, S selects a random string $W_i^{b_i}$ of length $2k(n)$, and a random bit c_i. (Recall that b_i denotes the value of the i-th wire induced by the input x. The simulator, of course, does not know this value.) For an input wire i, the simulator outputs $W_i^{b_i} \circ c_i$. For a gate t with input wires i, j and output wires $y_1, \ldots, y_m$ the simulator computes the active label $Q_t^{c_i, c_j} = E_{W_i^{b_i, c_j} \oplus W_j^{b_j, c_i}}(W_{y_1}^{b_{y_1}} \circ c_{y_1} \circ \cdots \circ W_{y_m}^{b_{y_m}} \circ c_{y_m})$ and sets the other three inactive labels of this gate to be encryptions of all-zeros strings of appropriate length under random keys; that is, for every two bits $(a_i, a_j) \neq (c_i, c_j)$, the simulator chooses uniformly a $k(n)$-bit string R_{a_i, a_j} and outputs $Q_l^{a_i, a_j} = E_{R_{a_i, a_j}}(0^{|W_{y_1}^{b_{y_1}} \circ c_{y_1} \circ \cdots \circ W_{y_m}^{b_{y_m}} \circ c_{y_m}|})$. Finally, for an output wire i, the simulator outputs $r_i = c_i - b_i$ (recall that b_i is known since $f_n(x)$ is given).

Since C_n can be constructed in polynomial time and since the encryption algorithm runs in polynomial time the simulator is also a polynomial-time algorithm. We refer to the gate labels constructed by the simulator as "fake" gate labels and to gate labels of $\hat{f}_n$ as "real" gate labels.

Assume, towards a contradiction, that there exists a (non-uniform) polynomial-size circuit family $\{A_n\}$, a polynomial $p(\cdot)$, a string family $\{x_n\}$, $|x_n| = n$, such that for infinitely many n's it holds that

$$\Delta(n) \overset{\text{def}}{=} \left| \Pr\left[A_n\left(S\left(1^n, f_n(x_n)\right)\right) = 1 \right] - \Pr\left[A_n\left(\hat{f}_n(x_n, U_{\mu(n)})\right) = 1 \right] \right| > \frac{1}{p(n)}.$$

We use a hybrid argument to show that such a distinguisher can be used to break the encryption hence deriving a contradiction.

5.2.4.2 Hybrid Encodings

From now on we fix n and let $k = k(n)$. We construct a sequence of hybrid distributions that depend on x_n, and mix "real" gates labels and "fake" ones, such that one endpoint corresponds to the simulated output (in which all the gates have "fake" labels) and the other endpoint corresponds to $\hat{f}_n(x_n, U_{\mu(n)})$ (in which all the gates have real labels). Hence, if the extreme hybrids can be efficiently distinguished then there must be two neighboring hybrids that can be efficiently distinguished.

The Hybrids H_t^n   First, we order the gates of C_n in topological order. That is, if the gate t uses the output of gate t', then $t' < t$. Now, for every $t = 0, \ldots, \Gamma(n)$, we define the hybrid algorithm H_t^n that constructs "fake" labels for the first t gates and "real" labels for the rest of the gates:

1. For every wire i uniformly choose two $2k$-bit strings $W_i^{b_i}$, $W_i^{1-b_i}$ and a random bit c_i.
2. For every input wire i output $W_i^{b_i} \circ c_i$.
3. For every gate $t' \leq t$ with input wires i, j and output wires $y_1, \ldots, y_m$ output

$$Q_{t'}^{c_i, c_j} = E_{W_i^{b_i, c_j} \oplus W_j^{b_j, c_i}} \left(W_{y_1}^{b_{y_1}} \circ c_{y_1} \circ \cdots \circ W_{y_m}^{b_{y_m}} \circ c_{y_m} \right),$$

and for every choice of $(a_i, a_j) \in \{0, 1\}^2$ that is different from (c_i, c_j), uniformly choose a k-bit string R_{a_i, a_j} and output $Q_{t'}^{a_i, a_j} = E_{R_{a_i, a_j}} (0^{|W_{y_1}^{b_{y_1}} \circ c_{y_1} \circ \cdots \circ W_{y_m}^{b_{y_m}} \circ c_{y_m}|})$.
4. For every gate $t' > t$, let g be the function that t' computes (AND or OR), let i, j be the input wires of t' and let $y_1, \ldots, y_m$ be its output wires. Use x_n to compute the value of $b_i(x_n), b_j(x_n)$, and set $r_i = c_i - b_i$ and $r_j = c_j - b_j$. For every choice of $(a_i, a_j) \in \{0, 1\}^2$, compute $Q_{t'}^{a_i, a_j}$ exactly as in (5.1), and output it.
5. For every output wire i compute b_i and output $r_i = c_i - b_i$.

Claim 5.3   *There exist some $0 \leq t \leq \Gamma(n) - 1$ such that A_n distinguishes between H_t^n and H_{t+1}^n with advantage $\frac{\Delta(n)}{\Gamma(n)}$.*

Proof First, note that H_t^n uses the string x_n only when constructing real labels, that is in Step 4. Steps 1–3 can be performed without any knowledge on x_n, and Step 5 requires only the knowledge of $f_n(x_n)$. Obviously, the algorithm $H_{\Gamma(n)}^n$ is just a different description of the simulator S, and therefore $S(1^n, f_n(x_n)) \equiv H_{\Gamma(n)}^n$. We also claim that the second extreme hybrid, H_0^n, coincides with the distribution of the "real" encoding, $\hat{f}_n(x_n, U_{\mu(n)})$. To see this note that (1) the strings W_i^0, W_i^1 are chosen uniformly and independently by H_0^n, as they are in $\hat{f}_n(x_n, U_{\mu(n)})$; and (2) since H_0^n chooses the c_i's uniformly and independently and sets $r_i = c_i - b_i$ then the r_i's themselves are also distributed uniformly and independently exactly as they are in $\hat{f}_n(x_n, U_{\mu(n)})$. Since for every gate t the value of $Q_t^{a_i, a_j}$ is a function of the random variables, and since it is computed by H_0^n in the same way as in $\hat{f}_n(x_n, U_{\mu(n)})$, we get that $H_0^n \equiv \hat{f}_n(x_n, U_{\mu(n)})$.

Hence, we can write

$$\Delta(n) = \left| \Pr\left[A_n\left(H_0^n\right) = 1\right] - \Pr\left[A_n\left(H_{\Gamma(n)}^n\right) = 1\right] \right|$$

$$\leq \sum_{t=0}^{\Gamma(n)-1} \left| \Pr\left[A_n\left(H_t^n\right) = 1\right] - \Pr\left[A_n\left(H_{t+1}^n\right) = 1\right] \right|,$$

and so there exists some $0 \leq t \leq \Gamma(n) - 1$ such that A_n distinguishes between H_t^n and H_{t+1}^n with advantage $\frac{\Delta(n)}{\Gamma(n)}$.   □

5.2.4.3 Distinguishing Fake Gates from Real Gates

We now show that distinguishing H_t^n and H_{t+1}^n allows us to distinguish whether a single gate is real or fake. To do this, we define two random experiments $P_n(0)$ and $P_n(1)$, that produce a real gate and a fake gate, correspondingly.

The Experiments $P_n(0), P_n(1)$ Let i, j be the input wires of the gate t, let $y_1, \ldots, y_m$ be the output wires of t, let g be the function that gate t computes, and let $b_i, b_j, b_{y_1}, \ldots, b_{y_m}$ be the values of the corresponding wires induced by the input x_n. For $\sigma \in \{0, 1\}$, define the distribution $P_n(\sigma)$ as the output distribution of the following random process:

- Uniformly choose the $2k$-bit strings

$$W_i^{b_i}, W_i^{1-b_i}, W_j^{b_j}, W_j^{1-b_j}, W_{y_1}^{b_{y_1}}, W_{y_1}^{1-b_{y_1}}, \ldots, W_{y_m}^{b_{y_m}}, W_{y_m}^{1-b_{y_m}},$$

and the random bits $c_i, c_j, c_{y_1}, \ldots, c_{y_m}$.
- If $\sigma = 0$ then set $Q_t^{c_i, c_j}$ and the other three $Q_t^{a_i, a_j}$ exactly as in Step 3 of H_t^n.
- If $\sigma = 1$ then set $Q_t^{a_i, a_j}$ exactly as in Step 4 of H_t^n; that is, set

$$r_i = c_i - b_i, \qquad r_j = c_j - b_j,$$

and for every choice of $(a_i, a_j) \in \{0, 1\}^2$, let

$$Q_t^{a_i, a_j} = E_{W_i^{a_i - r_i, a_j} \oplus W_j^{a_j - r_j, a_i}} \left(W_{y_1}^{g(a_i - r_i, a_j - r_j)} \circ \left(g(a_i - r_i, a_j - r_j) + r_{y_1} \right) \circ \cdots \right.$$

$$\left. \circ W_{y_m}^{g(a_i - r_i, a_j - r_j)} \circ \left(g(a_i - r_i, a_j - r_j) + r_{y_m} \right) \right).$$

- Output the tuple

$$\left(W_i^{b_i}, W_j^{b_j}, W_{y_1}^{b_{y_1}}, W_{y_1}^{1-b_{y_1}}, \ldots, W_{y_m}^{b_{y_m}}, W_{y_m}^{1-b_{y_m}}, \right.$$

$$\left. c_i, c_j, c_{y_1}, \ldots, c_{y_m}, Q_t^{0,0}, Q_t^{0,1}, Q_t^{1,0}, Q_t^{1,1} \right).$$

Claim 5.4 *There exists a polynomial-size circuit A'_n that distinguishes between $P_n(0)$ and $P_n(1)$ with advantage $\frac{\Delta(n)}{\Gamma(n)}$.*

Proof The adversary A'_n uses the output of $P_n(\sigma)$ to construct one of the hybrids H^n_t and H^n_{t+1}, and then uses A_n to distinguish between them. Namely, given the output of P_n, the distinguisher A'_n invokes the algorithm H^n_t where the values of $(W^{b_i}_i, W^{b_j}_j, W^{b_{y_1}}_{y_1}, W^{1-b_{y_1}}_{y_1}, \ldots, W^{b_{y_m}}_{y_m}, W^{1-b_{y_m}}_{y_m}, c_i, c_j, c_{y_1}, \ldots, c_{y_m}, Q^{0,0}_t, Q^{0,1}_t, Q^{1,0}_t, Q^{1,1}_t)$ are set to the values given by P_n. By the definition of P_n, when $P_n(0)$ is invoked we get the distribution of H^n_t, that is the gate t is "fake"; on the other hand, if $P_n(1)$ is invoked then the gate t is "real" and we get the distribution of H^n_{t+1}. Hence, by Claim 5.3, A'_n has the desired advantage. Finally, since H^n_t runs in polynomial time (when x_n is given), the size of A'_n is indeed polynomial. $\square$

A Delicate Point   Note that P_n does not output the inactive keys of the wires i and j (which is crucial for Claim 5.5 to hold). However, the hybrid distributions use inactive keys of wires that either enter a real gate or leave a real gate (in the first case the inactive keys are used as the keys of the gate label encryption whereas in the latter case the inactive keys are being encrypted). Hence, we do not need these inactive keys to construct the rest of the distribution H^n_t (or H^n_{t+1}), as i and j are output wires of gates that precede t and therefore are "fake" gates. This is the reason why we had to sort the gates. On the other hand, the process P_n must output the inactive keys of the output wires of the gate $y_1, \ldots, y_m$, since these wires might enter as inputs to another gate $t' > t$ which is a "real" gate in both H^n_t and H^n_{t+1}.

5.2.4.4 Deriving a Contradiction

We now define a related experiment P'_n in which some of the randomness used by P_n is fixed. Specifically, we fix the random strings $W^{b_{y_1}}_{y_1}, W^{1-b_{y_1}}_{y_1}, \ldots, W^{b_{y_m}}_{y_m}, W^{1-b_{y_m}}_{y_m}, c_i, c_j, c_{y_1}, \ldots, c_{y_m}$ to some value such that the resulting experiments still can be distinguished by A'_n with advantage $\frac{\Delta(n)}{\Gamma(n)}$. (The existence of such strings is promised by an averaging argument.) For simplicity, we omit the fixed strings from the output of this new experiment. The experiments $P'_n(0)$ and $P'_n(1)$ can still be distinguished by some polynomial-size circuit with advantage $\frac{\Delta(n)}{\Gamma(n)}$. (Such a distinguisher can be constructed by incorporating the omitted fixed strings into A'_n.) Hence, by the contradiction hypothesis, it follows that this advantage is greater than $\frac{1}{\Gamma(n)p(n)}$ for infinitely many n's. As $\Gamma(n)$ is polynomial in n (since C_n is of polynomial size) we deduce that the distribution ensembles $\{P'_n(0)\}_{n \in \mathbb{N}}$ and $\{P'_n(1)\}_{n \in \mathbb{N}}$ are not computationally indistinguishable, in contradiction with the following claim.

Claim 5.5 $\{P'_n(0)\}_{n \in \mathbb{N}} \overset{c}{\equiv} \{P'_n(1)\}_{n \in \mathbb{N}}$.

Proof Fix some n. For both distributions $P'_n(0)$ and $P'_n(1)$, the first two entries $(W^{b_i}_i, W^{b_j}_j)$ are two uniformly and independently $2k(n)$-length strings, and the active label Q^{b_i, b_j}_t is a function of $(W^{b_i}_i, W^{b_j}_j)$ and the fixed strings. Hence, the distributions $P'_n(0)$ and $P'_n(1)$ differ only in the inactive labels Q^{a_i, a_j}_t for

$(a_i, a_j) \neq (c_i, c_j)$. Hence, we would like to show that, conditioned on any fixing of $(W_i^{b_i}, W_j^{b_j})$, the tuple $(Q_t^{a_i,a_j})_{(a_i,a_j) \neq (c_i,c_j)}$ as computed by $P_n'(0)$ is computationally indistinguishable from the tuple $(Q_t^{a_i,a_j})_{(a_i,a_j) \neq (c_i,c_j)}$ as computed by $P_n'(1)$. In order to prove this, we will take a closer look at these distributions conditioned on some fixed value for $(W_i^{b_i}, W_j^{b_j})$.

First, recall that in $P_n'(0)$ each of the entries of $(Q_t^{a_i,a_j})_{(a_i,a_j) \neq (c_i,c_j)}$ is an all-zeros string that was encrypted under uniformly and independently chosen key R_{a_i,a_j}. In the second distribution $P_n'(1)$, the entry $Q_t^{a_i,a_j}$ is computed by encrypting a "meaningful" fixed message (as we fixed the c_i's and W_y's) under the key $K^{a_i,a_j} = W_i^{a_i-r_i,a_j} \oplus W_j^{a_j-r_j,a_i}$. We will show the following.

Claim 5.6 *The keys* $(K^{a_i,a_j})_{(a_i,a_j) \neq (c_i,c_j)}$ *are uniformly and independently distributed.*

Proof Fix $(a_i, a_j) \neq (c_i, c_j)$. We make the following observations: (1) Each of the strings $W_i^{a_i-r_i,a_j}, W_j^{a_j-r_j,a_i}$ is chosen uniformly; (2) $K^{a_i,a_j} = W_i^{a_i-r_i,a_j} \oplus W_j^{a_j-r_j,a_i}$ but each of the strings $W_i^{a_i-r_i,a_j}, W_j^{a_j-r_j,a_i}$ is independent of the other two keys $(K^{a_i',a_j'})_{(a_i',a_j') \neq (a_i,a_j),(c_i,c_j)}$. (3) At least one of the strings $W_i^{a_i-r_i,a_j}$, $W_j^{a_j-r_j,a_i}$ is independent of $W_i^{b_i}, W_j^{b_j}$ since $(a_i, a_j) \neq (c_i, c_j)$; It follows that, even when we condition on $W_i^{a_i-r_i,a_j}, W_j^{a_j-r_j,a_i}$, each entry of $(K^{a_i,a_j})_{(a_i,a_j) \neq (c_i,c_j)}$ is distributed uniformly and independently of the other entries. $\qquad \square$

So in both experiments $P_n'(1)$ and $P_n'(0)$ (conditioned on $W_i^{b_i}, W_j^{b_j}$), the triple of inactive labels $(Q_t^{a_i,a_j})_{(a_i,a_j) \neq (c_i,c_j)}$ is computed by encrypting a triple of fixed messages of length $p(n)$ under a triple of randomly chosen independent keys of length $k(n)$. (Recall that $k(n)$ is polynomial in n by definition, and $p(n) = O(|C_n|k(n)) = \text{poly}(n)$). Hence, by the security of the encryption scheme combined with Fact 2.7, it follows that the ensemble $(Q_t^{a_i,a_j})_{(a_i,a_j) \neq (c_i,c_j)}$ produced by $P_n'(1)$ is computationally indistinguishable from the corresponding ensemble produced by $P_n'(0)$, which completes the proof of Claim 5.5. $\qquad \square$

5.3 Applications

5.3.1 Relaxed Assumptions for Cryptography in NC⁰

In Chap. 4 we showed that computational randomized encoding preserves the security of many cryptographic primitives.[9] It follows from Theorem 5.2 that, under the

---

[9]In some cases, we will need to rely on *perfect* correctness, which we get "for free" in our main construction. See Table 4.2.

EOWF assumption, any such primitive can be computed in $\mathbf{NC}^0$ *if it exists at all* (i.e., can be computed in polynomial time). Formally, we have the following.

Theorem 5.3 *Suppose that the EOWF assumption holds. Then,*

1. *If there exists a public-key encryption scheme (resp., NIZK with an efficient prover or constant-round ZK proof with an efficient prover for every* **NP** *relation), then there exists such a scheme in which the encryption (prover) algorithm is in* $\mathbf{NC}_4^0$.
2. *If there exists a non-interactive commitment scheme, then there exists such a scheme in which the sender is in* $\mathbf{NC}_4^0$ *and the receiver is in* $\mathrm{AND}_n \circ \mathbf{NC}^0$.
3. *There exists a stateless symmetric encryption scheme (resp., digital signature, MAC, a constant-round ZK argument for every language in* **NP***) in which the encryption (signing, prover) algorithm is in* $\mathbf{NC}_4^0$.
4. *There exists a constant-round commitment scheme in which the sender is in* $\mathbf{NC}_4^0$ *and the receiver is in* $\mathrm{AND}_n \circ \mathbf{NC}^0$.
5. *For every polynomial-time computable function we have a (non-adaptive single-oracle) IHS in which the user is in* $\mathbf{NC}_5^0$ *and the oracle is in* **BPP**.

The existence of (stateless) symmetric encryption, signature, MAC, constant-round commitment scheme and constant-round ZK arguments for **NP**, does not require any additional assumption other than EOWF. This is a consequence of the fact that they all can be constructed (in polynomial time) from a OWF (see [70, 72]). For these primitives, we obtain more general (unconditional) results in the next subsection.

Remark 5.4 Theorem 5.3 reveals an interesting phenomenon. It appears that several cryptographic primitives (e.g., symmetric encryption schemes, digital signatures and MACs) can be implemented in $\mathbf{NC}^0$ despite the fact that their standard constructions rely on pseudorandom functions (PRFs) [74], which cannot be computed even in $\mathbf{AC}^0$ [108]. For such primitives, we actually construct a sequential PRF from the PRG (as in [74]), use it as a building block to obtain a sequential construction of the desired primitive (e.g., symmetric encryption), and finally reduce the parallel-time complexity of the resulting function using our machinery. Of course, the security of the PRF primitive itself is not inherited by its computational (or even perfect) encoding.

Remark 5.5 (**Concrete assumptions**) Theorem 5.3 allows an $\mathbf{NC}^0$ implementations of many primitives under several (new) concrete assumptions. In particular, since the EOWF assumption is implied by the intractability of factoring, the discrete logarithm problem, and lattice problems (see Remark 4.1), we can use any of these assumptions to obtain a digital signature (and MAC) whose signing algorithm is in $\mathbf{NC}_4^0$, as well as a non-interactive commitment scheme in which the sender is in $\mathbf{NC}_4^0$ and the receiver is in $\mathrm{AND}_n \circ \mathbf{NC}^0$. Recall that in Sect. 4.7.2 we showed that, under the same assumptions, there exists a two-round statistically hiding commitment in which the sender is in $\mathbf{NC}_4^0$ and the receiver is in $\mathrm{AND}_n \circ \mathbf{NC}^0$. Hence, by combining the two aforementioned commitments with the ZK protocol of [75] (as explained in Sect. 4.7.3) we get a constant-round ZK-proof for **NP** between an $\mathbf{AC}^0$ prover and an $\mathbf{AC}^0$ verifier assuming any of the above assumptions. We can also use Theorem 5.3 to obtain a NIZK with an $\mathbf{NC}^0_4$ prover for **NP** under the intractability of factoring. (This follows from the NIZK construction of [59] which can be based on the intractability of factoring [26, 72, Sect. C.4.1].)

5.3.1.1 Parallelizing the Receiver

As mentioned above, the computational encoding promised by Theorem 5.2 does not support parallel decoding. Thus, we get primitives in which the sender (i.e., the encrypting party, committing party, signer or prover, according to the case) is in $\mathbf{NC}^0$ but the receiver (the decrypting party or verifier) is not known to be in **NC**, even if we started with a primitive that has an **NC** receiver. The following theorem tries to partially remedy this state of affairs. Assuming the existence of a PRG with a good stretch in $\mathbf{NC}^1$, we can rely on Claim 5.2 to convert sender-receiver schemes in which both the receiver and the sender are in **NC** to ones in which the sender is in $\mathbf{NC}^0$ and the receiver is still in $\mathbf{NC}$.[10]

Theorem 5.4 *Let $\mathscr{X} = (G, S, R)$ be a sender-receiver cryptographic scheme whose security is respected by computational encoding (e.g., encryption, signature, MAC, commitment scheme, NIZK), where G is a key-generation algorithm (in case the scheme has one), $S \in \mathbf{NC}^s$ is the algorithm of the sender and $R \in \mathbf{NC}^r$ is the algorithm of the receiver. Then:*

- *If there exists a polynomial-stretch PRG in $\mathbf{NC}^1$, then there exists a similar scheme $\hat{\mathscr{X}} = (G, \hat{S}, \hat{R})$ in which $\hat{S} \in \mathbf{NC}^0$ and $\hat{R} \in \mathbf{NC}^{\max\{s+1,r\}}$.*
- *If there exists a linear-stretch PRG in $\mathbf{NC}^1$, then there exists a similar scheme $\hat{\mathscr{X}} = (G, \hat{S}, \hat{R})$, in which $\hat{S} \in \mathbf{NC}^0$ and $\hat{R} \in \mathbf{NC}^{\max\{s+2,r\}}$.*

Proof If there exists a polynomial-stretch (resp. linear-stretch) PRG in $\mathbf{NC}^1$, then we can use Claim 5.2 and get a computational encoding $\hat{S}$ for S in $\mathbf{NC}^0$ whose decoder B is in $\mathbf{NC}^{s+1}$ (resp. $\mathbf{NC}^{s+2}$). As usual, the new receiver $\hat{R}$ uses B to decode the encoding, and then applies the original receiver R to the result. Thus, $\hat{R}$ is in $\mathbf{NC}^{\max\{s+1,r\}}$ (resp. $\mathbf{NC}^{\max\{s+2,r\}}$). □

5.3.2 Parallel Reductions Between Cryptographic Primitives

In the previous section we showed that many cryptographic tasks can be performed in $\mathbf{NC}^0$ if they can be performed at all, relying on the assumption that an easy OWF

---

[10]Similarly, assuming a linear-stretch PRG in $\mathbf{NC}^1$, we can obtain, for every **NC** function, a (non-adaptive single-oracle) IHS in which the user is in $\mathbf{NC}^0$ and the oracle is in **NC**.

exists. Although EOWF is a very reasonable assumption, it is natural to ask what types of parallel reductions between primitives can be guaranteed *unconditionally*. In particular, such reductions would have consequences even if there exists a OWF in, say, $\mathbf{NC}^4$.

In this section, we consider the types of unconditional reductions that can be obtained using the machinery of Sect. 5.2. We focus on primitives that can be reduced to a OWF. We argue that for any such primitive $\mathscr{P}$, its polynomial-time reduction to a OWF can be collapsed into an $\mathbf{NC}^0$-reduction to a OWF. More specifically, we present an efficient "compiler" that takes the code of an arbitrary OWF F and outputs a description of an $\mathbf{NC}^0$ circuit C, having oracle access to a function F', such that for any OWF F' the circuit $C[F']$ implements $\mathscr{P}$.

A compiler as above proceeds as follows. Given the code of F, it first constructs a *code* for an efficient implementation P of $\mathscr{P}$. (In case we are given an efficient *black-box* reduction from $\mathscr{P}$ to a OWF, this code is obtained by plugging the code of F into this reduction.) Then, applying a constructive form of Theorem 5.1 to the code of P, the compiler obtains a code $\hat{P}$ of an $\mathbf{NC}^0$ circuit which implements $\mathscr{P}$ by making an oracle access to a OWF. This code of $\hat{P}$ defines the required $\mathbf{NC}^0$ reduction from $\mathscr{P}$ to a OWF. Note that the specification of $\hat{P}$ depends on the code of the given OWF F. Thus, the reduction makes a *non-black-box* use of the given OWF F, even if the polynomial-time reduction it is based on is fully black-box.

Based on the above we can obtain the following informal "meta-theorem".

Meta-Theorem 5.1 *Let $\mathscr{P}$ be a cryptographic primitive whose security is respected by computational encoding. Suppose that $\mathscr{F}$ is polynomial-time reducible to a OWF. Then, $\mathscr{P}$ is $\mathbf{NC}^0$-reducible to a OWF.*

Instantiating $\mathscr{F}$ by concrete primitives, we get the following corollary.

Corollary 5.2 *Let F be a OWF. Then:*

- *There exists a stateless symmetric encryption scheme (resp., digital signature or MAC) in which the encryption (signing) algorithm is in $\mathbf{NC}^0[F]$.*
- *There exists a constant-round commitment scheme (resp., constant-round coin-flipping protocol) in which the sender (first party) is in $\mathbf{NC}^0[F]$ and the receiver (second party) is in $\mathrm{AND}_n \circ \mathbf{NC}^0[F]$.*
- *For every $\mathbf{NP}$ language, there exists a constant-round ZK argument in which the prover is in $\mathbf{NC}^0[F]$.*

An illustration of the above appears in Fig. 1.1. Note that items 3, 4 of Theorem 5.3 can be derived from the above corollary, up to the exact locality.

5.3.2.1 Comparison with Known Reductions

The above results can be used to improve the parallel complexity of some known reductions. For example, Naor [114] shows a commitment scheme in which the sender is in $\mathbf{NC}^0[LG]$, where LG is a *linear-stretch* PRG. By using his construction, we derive a commitment scheme in which the sender (respectively, the receiver) is in $\mathbf{NC}^0[F]$ (respectively, $\mathrm{AND}_n \circ \mathbf{NC}^0[F]$) where F is a OWF. Since it is not known how to reduce a linear-stretch PRG to a OWF (or to a minimal PRG) even in $\mathbf{NC}$, we get a nontrivial parallel reduction.

Other interesting examples arise in the case of primitives that are based on PRFs, such as MACs, symmetric encryption, and identification (see [72, 74, 116] for these and other applications of PRFs). Since the known construction of a PRF from a PRG is sequential [74], it was not known how to reduce these primitives in parallel to a OWF, or even to a polynomial-stretch PRG (PPRG).[11] This fact motivated the study of parallel constructions of PRFs in [116, 117]. In particular, Naor and Reingold [116] introduce a new cryptographic primitive called a synthesizer (SYNTH), and show that PRFs can be implemented in $\mathbf{NC}^1[\mathrm{SYNTH}]$. This gives an $\mathbf{NC}^1$-reduction from cryptographic primitives such as symmetric encryption to synthesizers. By Corollary 5.2, we get that these primitives are in fact $\mathbf{NC}^0$-reducible to a OWF. Under parallel-reductions it seems that synthesizers are much stronger tha OWFs. (In fact, synthesizers seem even stronger than PPRGs, as $\mathrm{PPRG} \in \mathbf{NC}^0[\mathrm{SYNTH}]$ while we have no $\mathbf{NC}$-reduction from synthesizers to PPRG.) Hence, our results improve both the complexity of the reduction and the underlying assumption. It should be noted, however, that our reduction only improves the parallel-time complexity of the encrypting party, while the constructions of [116] yield $\mathbf{NC}^1$-reductions on both ends.

In contrast to the above, we show that a synthesizer in $\mathbf{NC}^i$ can be used to implement encryption in $\mathbf{NC}^i$ with decryption in $\mathbf{NC}$.[12] First, we use [116] to construct an encryption scheme (E, D) and a PPRG G such that E and D are in $\mathbf{NC}^1[\mathrm{SYNTH}] = \mathbf{NC}^{i+1}$ and G is in $\mathbf{NC}^0[\mathrm{SYNTH}] = \mathbf{NC}^i$. Next, by Claim 5.1 and Remark 5.2, we obtain an $\mathbf{NC}^0[G] = \mathbf{NC}^i$ computational encoding $\hat{E}$ for E whose decoder B is in $\mathbf{NC}^{2i}$. (We first use Claim 5.1 to construct one-time symmetric encryption (OE, OD) such that OE and OD are in $\mathbf{NC}^0[G] = \mathbf{NC}^i$. Then, we encode E by plugging OE into Construction 5.2 and obtain an $\mathbf{NC}^0[OE] = \mathbf{NC}^i$ computational encoding $\hat{E}$ for E. By Remark 5.2 the decoder B of $\hat{E}$ is in $\mathbf{NC}^i[OD] = \mathbf{NC}^{2i}$.) To decrypt ciphers of $\hat{E}$ we invoke the decoder B, and then apply the original decryption algorithm D to the result. Therefore, the decryption algorithm of our new scheme $\hat{D}$ is in $\mathbf{NC}^{\max(2i, i+1)}$.

5.3.2.2 Comparison with the Reductions of Chap. 4

Some $\mathbf{NC}^0$ reductions between cryptographic primitives were also obtained in Chap. 4 (see Remarks 4.5 and 4.6). In particular, there we considered the following

---

[11] Assuming that factoring is intractable (or, more generally, that there exists a OWF in **SREN**) it is provably impossible to obtain an $\mathbf{NC}^0$ reduction from PRFs to sublinear stretch PRGs or OWFs. See Sect. 4.8.

[12] For concreteness, we refer here only to the case of symmetric encryption, the case of other primitives which are $\mathbf{NC}^0$-reducible to a PRF (such as identification schemes and MACs) is analogous.

scenario: Let $\mathscr{G}$ be a primitive whose security is preserved by randomized encoding. Suppose that $G(x) = g(x, f(q_1(x)), \ldots, f(q_m(x)))$ defines a black-box construction of a primitive G of type $\mathscr{G}$ from a primitive f of type $\mathscr{F}$ where g is in **SREN** (or **PREN**) and the q_i's are in $\mathbf{NC}^0$. (The functions $g, q_1, \ldots, q_m$ are fixed by the reduction and do not depend on f.) Then, letting $\hat{g}((x, y_1, \ldots, y_m), r)$ be an $\mathbf{NC}^0$ encoding of g, the function $\hat{G}(x, r) = \hat{g}((x, f(q_1(x)), \ldots, f(q_m(x))), r)$ encodes G, and hence defines a (non-adaptive) black-box $\mathbf{NC}^0$ reduction from $\mathscr{G}$ to $\mathscr{F}$. An important example for such a reduction is the transformation of OWF to PRG from [83] (see Remark 4.5). In this case, the transformation is in $\mathbf{NC}^1$ and thus is improved to be in $\mathbf{NC}^0$.[13] Unlike these previous reductions, the current results are not restricted to non-adaptive black-box reductions which are computable in **SREN**. However, the reductions of this chapter are inherently non black-box and their structure depends on the code implementing the given oracle.

5.3.3 Secure Multiparty Computation

Secure multiparty computation (MPC) allows several parties to evaluate a function of their inputs in a distributed way, so that both the privacy of their inputs and the correctness of the outputs are maintained. These properties should hold, to the extent possible, even in the presence of an adversary who may corrupt at most t parties. This is typically formalized by comparing the adversary's interaction with the *real process*, in which the uncorrupted parties run the specified protocol on their inputs, with an ideal function evaluation process in which a trusted party is employed. The protocol is said to be *secure* if whatever the adversary "achieves" in the real process it could have also achieved by corrupting the ideal process. A bit more precisely, it is required that for every adversary A interacting with the real process there is an adversary A' interacting with the ideal process, such that outputs of these two interactions are *indistinguishable* from the point of view of an external environment. See, for example, [41, 42, 72], for more detailed and concrete definitions.

There is a variety of different models for secure computation. These models differ in the power of the adversary, the network structure, and the type of "environment" that tries to distinguish between the real process and the ideal process. In the *information-theoretic* setting, both the adversary and the distinguishing environment may be computationally unbounded, whereas in the *computational* setting they are both bounded to probabilistic polynomial time.

The notion of randomizing polynomials was originally motivated by the goal of minimizing the round complexity of MPC. The motivating observation of [92] was that the round complexity of most general protocols from the literature (e.g., those of [27, 45, 78]) is related to the algebraic *degree* of the function being computed.

---

[13] Similar examples are the $\mathbf{NC}^1$ transformation of one-to-one OWF to non-interactive commitment scheme (cf. [34]) and of distributionally OWF into standard OWF (cf. [90]).

Thus, by reducing the task of securely computing f to that of securely computing some related low-degree function, one can obtain round-efficient protocols for f.

Randomizing polynomials (or low-degree randomized encodings) provide precisely this type of reduction. More specifically, suppose that the input x to f is distributed between the parties, who wish to all learn the output $f(x)$. If f is represented by a vector $\hat{f}(x, r)$ of degree-d randomizing polynomials, then the secure computation of f can be *non-interactively* reduced to that of $\hat{f}$, where the latter is viewed as a *randomized* function of x. This reduction only requires each party to invoke the decoder of $\hat{f}$ on its local output, obtaining the corresponding output of f. The secure computation of $\hat{f}$, in turn, can be non-interactively reduced to that of a related *deterministic* function $\hat{f}'$ of the same degree d. The idea is to let $\hat{f}'(x, r^1, \ldots, r^{t+1}) \stackrel{\text{def}}{=} p(x, r^1 \oplus \cdots \oplus r^{t+1})$ (where t is a bound on the number of corrupted parties), assign each input vector r^j to a distinct player, and instruct it to pick it at random. (See [92] for more details.) This second reduction step is also non-interactive. Thus, any secure protocol for $\hat{f}'$ or $\hat{f}$ gives rise to a secure protocol for f with the same number of rounds. The non-interactive nature of the reduction makes it insensitive to almost all aspects of the security model.

Previous constructions of (perfect or statistical) randomizing polynomials [49, 92, 93] provided *information-theoretic* reductions of the type discussed above. In particular, if the protocol used for evaluating $\hat{f}'$ is information-theoretically secure, then so is the resulting protocol for f. The main limitation of these previous reductions is that they efficiently apply only to restricted classes of functions, typically related to different log-space classes. This situation is remedied in the current work, where we obtain (under the EOWF assumption) a *general* secure reduction from a function f to a related degree-3 function $\hat{f}'$. The main price we pay is that the security of the reduction is no longer information-theoretic. Thus, even if the underlying protocol for $\hat{f}'$ is secure in the information-theoretic sense, the resulting protocol for f will only be computationally secure.

To formulate the above we need the following definitions.

Definition 5.3 (Secure computation) Let $f(x_1, \ldots, x_n)$ be an m-party functionality, i.e., a (possibly randomized) mapping from m inputs of equal length into m outputs. Let π be an m-party protocol. We formulate the requirement that π securely computes f by comparing the following "real process" and "ideal process".

The real process A *t-bounded* adversary A attacking the *real process* is a probabilistic polynomial-time algorithm, who may corrupt up to t parties and observe all of their internal data. At the end of the interaction, the adversary may output an arbitrary function of its view, which consists of the inputs, the random coin tosses, and the incoming messages of the corrupted parties. We distinguish between *passive* vs. *active* adversaries and between *adaptive* vs. *non-adaptive* adversaries. If the adversary is active, it has full control over the messages sent by the corrupted parties, whereas if it is passive, it follows the protocol's instructions (but may try to deduce information by performing computations on observed data). When the set of corrupted parties has to be chosen in advance, we say that the adversary is non-adaptive, and otherwise say that it is adaptive. Given an m-tuple of inputs $(x_1, \ldots, x_m) \in (\{0,1\}^n)^m$, the *output of the real process* is defined as the random variable containing the *concatenation* of the adversary's output with the outputs and identities of the uncorrupted parties. We denote this output by $\mathrm{REAL}_{\pi,A}(x_1, \ldots, x_m)$.

The ideal process In the ideal process, an incorruptible trusted party is employed for computing the given functionality. That is, the "protocol" in the ideal process instructs each party to send its input to the trusted party, who computes the functionality f and sends to each party its output. The interaction of a t-bounded adversary A' with the ideal process and the output of the ideal process are defined analogously to the above definitions for the real process. The adversary attacking the ideal process will also be referred to as a *simulator*. We denote the output of the ideal process on the inputs $(x_1, \ldots, x_m) \in (\{0,1\}^n)^m$ by $\mathrm{IDEAL}_{f,A'}(x_1, \ldots, x_m)$.

The protocol π is said to t-*securely* realize the given functionality f with respect to a specified type of adversary (namely, passive or active, adaptive or non-adaptive) if for any probabilistic polynomial-time t-bounded adversary A attacking the real process, there exists a probabilistic polynomial-time t-bounded simulator A' attacking the ideal process, such that for any sequence of m-tuples $\{\bar{x}_n\}$ such that $\bar{x}_n \in (\{0,1\}^n)^m$, it holds that $\mathrm{REAL}_{\pi,A}(\bar{x}_n) \overset{c}{\equiv} \mathrm{IDEAL}_{f,S}(\bar{x}_n)$.

Secure Reductions To define secure reductions, consider the following *hybrid* model. An m-party protocol augmented with an oracle to the m-party functionality g is a standard protocol in which the parties are allowed to invoke g, i.e., a trusted party to which they can securely send inputs and receive the corresponding outputs. The notion of t-security generalizes to protocols augmented with an oracle in the natural way.

Definition 5.4 Let f and g be m-party functionalities. A t-*secure* reduction from f to g is an m-party protocol that given an oracle access to the functionality g, t-securely realizes the functionality f (with respect to a specified type of adversary). We say that the reduction is *non-interactive* if it involves a single call to f (and possibly local computations on inputs and outputs), but no further communication.

Appropriate composition theorems, e.g., [72, Theorems 7.3.3, 7.4.3], guarantee that the call to g can be replaced by any secure protocol realizing g, without violating the security of the high-level protocol for f.[14] Using the above terminology, Theorem 5.2 has the following corollary.

---

[14] Actually, for the composition theorem to go through, Definition 5.3 should be augmented by providing players and adversaries with auxiliary inputs. We ignore this technicality here, and note that the results in this section apply (with essentially the same proofs) to the augmented model as well.

Theorem 5.5 *Suppose the EOWF assumption holds. Let $f(x_1, \ldots, x_m)$ be an m-party functionality computed by a (uniform) circuit family of size $s(n)$. Then, for any $\varepsilon > 0$, there is a non-interactive, computationally $(m - 1)$-secure reduction from f to either of the following two efficient functionalities:*

- *a randomized functionality $\hat{f}(x_1, \ldots, x_m)$ of degree 3 (over $\mathbb{F}_2$) with a random input and output of length $O(s(n) \cdot n^\varepsilon)$ each;*
- *a deterministic functionality $\hat{f}'(x_1', \ldots, x_m')$ of degree 3 (over $\mathbb{F}_2$) with input length $O(m \cdot s(n) \cdot n^\varepsilon)$ and output length $O(s(n) \cdot n^\varepsilon)$.*

Both reductions are non-interactive in the sense that they involve a single call to $\hat{f}$ or $\hat{f}'$ and no further interaction. They both apply regardless of whether the adversary is passive or active, adaptive or non-adaptive.

Proof The second item follows from the first via standard (non-interactive, degree-preserving) secure reduction from randomized functionalities to deterministic functionalities (see [72, Proposition 7.3.4]). Thus we will only prove the first item. Assume, without loss of generality, that f is a deterministic functionality that returns the same output to all the parties.[15] Let $\hat{f}(x, r)$ be the computational encoding of $f(x)$ promised by Theorem 5.2. (Recall that $\hat{f}$ is indeed a degree 3 function having $O(s(n) \cdot n^\varepsilon)$ random inputs and outputs.) The following protocol $(m - 1)$-securely reduces the computation of $f(x)$ to $\hat{f}(x, r)$ (where $\hat{f}$ is viewed as a randomized functionality whose randomness is r).

- Inputs: Party i gets input $x_i \in \{0, 1\}^n$.
- Party i invokes the (randomized) oracle $\hat{f}$ with query x_i, and receives an output $\hat{y}$.
- Outputs: Each party locally applies the decoder B of the encoding to the answer $\hat{y}$ received from the oracle, and outputs the result.

We start by showing that the reduction is $(m - 1)$-secure against a passive non-adaptive adversary. Let A be such an adversary that attacks some set $I \subset [m]$ of the players. Then, the output of the real process is $(A(x_I, \hat{f}(x, r)), B(\hat{f}(x, r)), \bar{I})$ where $x_I = (x_i)_{i \in I}$, $\bar{I} \stackrel{\text{def}}{=} [m] \setminus I$ and r is a uniformly chosen string of an appropriate length. We define a (passive non-adaptive) simulator A' that attacks the ideal process in the natural way: that is, $A'(x_I, y) = A(x_I, S(y))$, where y is the answer received from the trusted party (i.e., $f(x)$) and S is the computationally private simulator of the encoding. Thus, the output of the ideal process is $(A'(x_I, f(x)), f(x), \bar{I})$. By the definition of A', the privacy of the encoding $\hat{f}$ and Fact 2.8, we have,

---

[15] To handle randomized functionalities we use the non-interactive secure reduction mentioned above. Now, we can $(m - 1)$-securely reduce f to a single-output functionality by letting each party mask its output f_i with a private randomness. That is, $f'((x_1, r_1) \ldots, (x_m, r_m)) = ((f_1(x_1) \oplus r_1) \circ \cdots \circ (f_1(x_m) \oplus r_m))$. As both reductions are non-interactive the resulting reduction is also non-interactive. Moreover, the circuit size of f' is linear in the size of the circuit that computes the original function.

$$\text{IDEAL}(x) \equiv \left(A\big(x_I, S(f(x))\big), f(x), \bar{I}\right) \overset{c}{\equiv} \left(A\big(x_I, \hat{f}(x,r)\big), B\big(\hat{f}(x,r)\big), \bar{I}\right)$$
$$\equiv \text{REAL}(x),$$

which finishes the proof.

We now sketch the security proof for the case of an adversary A which is both adaptive and active. (The non-adaptive active case as well as the adaptive passive case are treated similarly.) An attack by A has the following form: (1) Before calling the oracle $\hat{f}$, in each step A may decide (according to his current view) to corrupt some party i and learn its input x_i. (2) When the oracle $\hat{f}$ is invoked A changes the input of each corrupted party i to some value x_i', which is handed to the $\hat{f}$ oracle. (3) After the parties call the oracle on some (partially corrupted) input $x' = (x_I', x_{\bar{I}})$, the oracle returns a randomized encoding $\hat{f}(x')$ to the adversary, and now A may adaptively corrupt additional parties. Finally, A outputs some function of its entire view. For every such adversary we construct a simulator A' that attacks the ideal process by invoking A and emulating his interaction with the real process. Namely: (1) before the call to the trusted party we let A choose (in each step) which party to corrupt and feed it with the input we learn; (2) when the trusted party is invoked we let A pick x_I' according to its current view and send these x_I' to the f oracle; (3) given the result $y = f(x_I', x_{\bar{I}})$ returned by the oracle, we invoke the simulator (of the encoding) on y and feed the result to A. Finally, we let A pick new corrupted players as in step (1). We claim that in each step of the protocol the view of A when interacting with the real process is computationally indistinguishable from the view of A when it is invoked by A' in the ideal process. (In fact, before the call to the oracle these views are identically distributed.) Hence, the outputs of the two processes are also computationally indistinguishable.   □

A high-level corollary of Theorem 5.5 is that computing arbitrary polynomial-time computable functionalities is as easy as computing degree-3 functionalities. Thus, when designing new MPC protocols, it suffices to consider degree-3 functionalities which are often easier to handle.

More concretely, Theorem 5.5 gives rise to new, conceptually simpler, constant-round protocols for general functionalities. For instance, a combination of this result with the "BGW protocol" [27] gives a simpler alternative to the constant-round protocol of Beaver, Micali, and Rogaway [25]. The resulting protocol will be more round-efficient, and in some cases (depending on the number of parties and the "easiness" of the OWF) even more communication-efficient than the protocol of [25]. Moreover, it was shown in [52] that Theorem 5.5 can be used in order to obtain the first constant-round protocol which is *scalable*, in the sense that the amortized work per player does not grow (and in some cases even vanishes) with the number of players. On the downside, Theorem 5.5 relies on a stronger assumption than the protocol from [25] (an easy OWF vs. an arbitrary OWF).

An interesting open question, which is motivated mainly from the point of view of the MPC application, is to come up with an "arithmetic" variant of the construction. That is, given an arithmetic circuit C, say with addition and multiplication gates, construct a vector of computationally private randomizing polynomials of size poly($|C|$) which makes a *black-box* use of the underlying field. The latter requirement means that the same polynomials should represent C over any field, ruling out the option of simulating arithmetic field operations by boolean operations. Such a result is known for weaker arithmetic models such as formulas and branching programs (see [49]). Recently, some progress has been made in extending this result to the case of general arithmetic circuits [19].

Chapter 6
One-Way Functions with Optimal Output Locality

Abstract In Chap. 4 it was shown that, under relatively mild assumptions, there exist one-way functions (OWFs) in $\mathbf{NC}_4^0$. This result is not far from optimal as there is no OWF in $\mathbf{NC}_2^0$. In this chapter we partially close this gap by providing an evidence for the existence of OWF in $\mathbf{NC}_3^0$. This is done in two steps: (1) we describe a new variant of randomized encoding that allows us to obtain a OWF in $\mathbf{NC}_3^0$ from a "robust" OWF which remains hard to invert even when some information on the preimage x is leaked; (2) we show how to construct a robust OWF assuming that a random function of (arbitrarily large) constant locality is one-way. (A similar assumption was previously made by Goldreich in Electron. Colloq. Comput. Complex. 7:090, 2000.)

6.1 Introduction

In Chaps. 4 and 5 we have established the existence of cryptography in $\mathbf{NC}_4^0$. On the other hand, we know that 2-local functions, being efficiently invertible, cannot be used for cryptography. This leaves open the possibility of 3-local cryptography. In this chapter we attempt to close this gap for the case of OWF, providing positive evidence for the existence of OWF in $\mathbf{NC}_3^0$.

A natural approach for closing the gap would be to reduce the degree of our general construction of randomized encodings from 3 to 2. (Indeed, Construction 4.1 transforms a degree-2 encoding into one in $\mathbf{NC}_3^0$.) However, the results of [92] provide some evidence against the prospects of this general approach, ruling out the existence of degree-2 perfectly private encodings for most nontrivial functions (see Sect. 3.3.3). Thus, we may take the following two alternative approaches: (1) employ a relaxed variant of randomized encodings which enables a degree-2 representation of general functions; and (2) seek *direct* constructions of degree-2 OWF based on specific intractability assumptions.

In Chap. 8, we will use approach (2) to construct a OWF (and even a PRG) with optimal locality based on the presumed intractability of decoding a random linear code. In this chapter we demonstrate the usefulness of approach (1) by employing degree-2 randomized encodings with a weak (but nontrivial) privacy property that we call *semi-privacy*. This encoding allows us to construct a OWF with optimal locality based on a OWF that enjoys a certain strong "robustness" property. We

B. Applebaum, *Cryptography in Constant Parallel Time*,
Information Security and Cryptography,
DOI 10.1007/978-3-642-17367-7_6, © Springer-Verlag Berlin Heidelberg 2014

also show how to construct such a robust OWF assuming that a random function of large constant locality is one-way (a similar assumption was suggested in [69]). We stress that our approach does not yield a general result in the spirit of the results of Chap. 4. Thus, we happen to pay for optimal degree and locality with the loss of generality.

6.1.1 Semi-private Randomized Encoding and Robust OWF

Let $\hat{f}$ be a randomized encoding of f. Recall that, according to the privacy property, the output distribution of $\hat{f}(x, r)$ (induced by a uniform choice of r) should hide all the information about x except for the value $f(x)$. Semi-privacy relaxes this requirement by insisting that the input x remain hidden by $\hat{f}(x, r)$ only in the case that $f(x)$ takes some specific value, say 0. (If $f(x)$ is different from this value, $\hat{f}(x, r)$ fully reveals x.) As it turns out, this relaxed privacy requirement is sufficiently liberal to allow a degree-2 encoding of general boolean functions.

Given any OWF f, one could attempt to apply a semi-private encoding as described above to every output bit of f, obtaining a degree-2 function $\hat{f}$. However, $\hat{f}$ will typically not be one-way: every output bit of f that evaluates to 1 might reveal the entire input. This motivates the following notion of a *robust* OWF. Loosely speaking, a OWF f is said to be robust if it remains (slightly) hard to invert even if a random subset of its output bits are "exposed", in the sense that all input bits leading to these outputs are revealed.

Formally, consider the following inversion game. First, we choose a random input $x \in \{0, 1\}^n$, compute $f(x)$ and send it to the adversary. Then, for each output bit of f we toss a coin b_i. If $b_i = 1$, we allow the adversary to see the bits of x that influence the i-th output bit. That is, we send $(x_{K(i)}, i, b_i)$ to the adversary, where $K(i) \subseteq [n]$ is the set of inputs that affects the i-th output bit. If $b_i = 0$, we reveal nothing regarding x and send (i, b_i) to the adversary. The adversary wins the game if she finds a preimage x' which is consistent with the game, i.e., $f(x') = f(x)$, and $x'_{K(i)} = x_{K(i)}$ whenever $b_i = 1$. The function is robust one-way if, for some polynomial $p(\cdot)$, any efficient adversary fails to find a consistent preimage with probability $1/p(n)$. (See Sect. 6.2 for a detailed definition.)

Intuitively, the purpose of the robustness requirement is to guarantee that the information leaked by the semi-private encoding leaves enough uncertainty about the input to make inversion difficult. Indeed, we show that when semi-private randomized encoding (SPRE) is applied to a (slightly modified) robust OWF the resulting function is distributionally one-way. Hence, a construction of degree-2 SPRE can be used to convert a robust OWF to a distributionally OWF with degree 2. Furthermore, it turns out that it is possible to convert the latter to a standard OWF with similar degree.

6.1.2 Constructing Robust OWF

We construct a robust OWF under the assumption that a random function f : $\{0, 1\}^n \rightarrow \{0, 1\}^n$ of large constant locality d is weakly one-way. Roughly speaking, the circuit that computes f is sampled by connecting each output to a random set of d inputs, and by placing at each output-gate a random predicate $P_i : \{0, 1\}^d \rightarrow \{0, 1\}$ where d is some (arbitrarily large) constant.[1] Then, we assume that, for some polynomial $p(\cdot)$, any efficient adversary that gets the description of f fails to invert it (on a randomly chosen input) with probability at least $1/p(n)$. Namely, we assume that the above procedure defines a collection of weakly OWFs. (See [70, Definition 2.4.3].) A similar assumption was proposed by Goldreich in [69], where it is shown that if the dependencies graph G satisfies some expansion property, then the function resists a natural class of inverting attacks. Since a random graph has good expansion with high probability, our variant resists the same class of attacks.

We show that, under the above assumption, a randomly chosen function h : $\{0, 1\}^{n \cdot e^d} \rightarrow \{0, 1\}^{2n}$ with (average) locality d/n is robust one-way. Intuitively, h is an extension of f to a bigger graph such that, with non-negligible probability, the random exposure of h, contains a copy of f; thus, it is slightly hard to invert and h is robust one-way.

Organization   Section 6.2 provides some preliminaries including a generalization of statistical randomized encoding. Semi-private randomized encodings are defined, constructed and analyzed in Sect. 6.3. In Sect. 6.4 we define the notion of robust one-way functions, and present a construction based on the hardness of inverting a random local function.

6.2 Preliminaries

We will need the following generalized form of randomized encoding.

Definition 6.1 (Generalized randomized encoding)   Let $h : \{0, 1\}^n \rightarrow \{0, 1\}^{l(n)}$ be an efficiently computable function. We say that $g : \{0, 1\}^n \times \{0, 1\}^m \rightarrow \{0, 1\}^{s(n)}$ is a *generalized statistical randomized encoding* (GRE) of h if there exists a function $f : \{0, 1\}^n \times \{0, 1\}^m \rightarrow \{0, 1\}^{s(n)}$ such that:

- f is a statistical randomized encoding of h.
- f is *isomorphic* to g in the sense that there exists permutation $\pi : \{0, 1\}^{n+m} \rightarrow \{0, 1\}^{n+m}$ such that for every $x \in \{0, 1\}^{n+m}$, $f(x) = g(\pi(x))$; and π is efficiently computable and efficiently invertible (in time poly(n)).

---

[1]More precisely, each input is connected to an output with probability d/n and so the *expected* degree is d.

It is not hard to show that GRE typically preserves cryptographic hardness. Specifically, we will need the following extension of Lemma 4.6, namely, if a function h is weakly one-way then its statistical generalized encoding g is distributionally one-way (see Definition 4.2).

Lemma 6.1 *If h is a weak-OWF and g is a GRE of h, then g is a distributional OWF.*

Proof Let f be a (standard) statistical encoding of h which is isomorphic to g via the isomorphism π. By Lemma 4.6, f is a distributional one-way function. We will show that g is a distributional one-way function as well. To simplify notation, we view f and g as deterministic functions that take a single input $x \in \{0, 1\}^n$ and map it into an $l(n)$-bit string.

Let A be an efficient adversary which distributionally inverts g with success $\varepsilon(n)$, i.e.,

$$\varepsilon(n) = \left\| \left(A(1^n, g(y)), g(y) \right) - \left(y, g(y) \right) \right\|,$$

where y is uniformly chosen from $\{0, 1\}^n$. Consider the adversary $\hat{A}$ which given $(1^n, z)$ invokes A on the same input, and translates the result y' to $x' = \pi^{-1}(y')$. It is not hard to see that $\hat{A}$ breaks f with the same advantage as A. Formally, we should show that

$$\left\| \left(\hat{A}(1^n, f(x)), f(x) \right) - \left(x, f(x) \right) \right\| \leq \varepsilon(n), \tag{6.1}$$

where x is uniformly chosen from $\{0, 1\}^n$. Let $z = f(x)$ and $y = \pi(x)$. Consider the mapping T which given a pair of strings $a \in \{0, 1\}^n, b \in \{0, 1\}^{l(n)}$ outputs $(\pi^{-1}(a), b)$. By Fact 2.3 and our assumption on A,

$$\left\| T\left(A(1^n, g(y)), g(y) \right) - T\left(y, g(y) \right) \right\| \leq \varepsilon(n).$$

However, by definition, $T(A(1^n, g(y)), g(y)) = (\hat{A}(1^n, f(x)), f(x))$ and, similarly, $T(y, g(y)) = (x, f(x))$, and so Eq. (6.1) follows. $\square$

The following lemma shows that in order to construct a degree-2 OWF, it suffices to find a degree-2 distributional OWF.

Lemma 6.2 *A degree-2 distributional OWF implies a degree-2 OWF in $\mathbf{NC}_3^0$.*

Proof First observe that the standard transformation from weak-OWF to (standard) OWF (cf. [144], Theorem 2.3.2 of [70]) preserves the algebraic degree. Therefore, a degree-2 weak OWF can be transformed into a degree-2 OWF. The latter can be converted to a degree-2 OWF in $\mathbf{NC}_3^0$ via Construction 4.1. Hence it is left to show how to transform a degree-2 distributional OWF into a degree-2 weak OWF.

Let f be a degree-2 distributional OWF. Consider the function

$$F(x, i, h) = \left(f(x), h_i(x), i, h \right),$$

where $x \in \{0,1\}^n$, $i \in \{1, \ldots, n\}$, $h : \{0,1\}^n \to \{0,1\}^n$ is a pairwise independent hash function, and h_i denotes the i-bit-long prefix of $h(x)$. This function was defined by Impagliazzo and Luby [90], who showed that in this case F is weakly one-way (see also [70], p. 96). Note that $h(x)$ can be computed as a degree-2 function of x and (the representation of) h by using the hash family $h_{M,v}(x) = Mx + v$, where M is an $n \times n$ matrix and v is a vector of length n. However, $h_i(x)$ is not of degree 2 when considered as a function of h, x and i, since "chopping" the last $n - i$ bits of $h(x)$ raises the degree of the function when i is not fixed.

We get around this problem by applying n copies of F on independent inputs, where each copy uses a different i. Namely, we define the function $F'((x^{(i)}, h^{(i)})_{i \in [n]}) \overset{\text{def}}{=} (F(x^{(i)}, i, h^{(i)}))_{i \in [n]}$. Since each of the i's is now fixed, the resulting function F' can be computed by degree-2 polynomials over $\mathbb{F}_2$. Moreover, it is not hard to verify that F' is weakly one-way if F is weakly one-way. We briefly sketch the argument. Given an efficient inverting algorithm A for F', one can invert $y = F(x, i, h) = (f(x), h_i(x), i, h)$ as follows. For every $j \neq i$, uniformly and independently choose $x^{(j)}, h^{(j)}$, set $z_j = F(x^{(j)}, j, h^{(j)})$ and $z_i = y$, then invoke A on $(z_j)_{j \in [n]}$ and output the i-th block of the answer. This inversion algorithm for F has the same success probability as A on a polynomially related input-length. $\square$

6.3 Semi-private Randomized Encoding

In this section we define the notion of semi-private randomized encoding (SPRE), describe a construction of degree-2 SPRE for functions with low (logarithmic) locality, and analyze the effect of encoding a function with SPRE. For simplicity, the results of this section are formulated in terms of functions, however they easily extend to the case of function ensembles.

6.3.1 Definition and Construction

The following definition relaxes the notion of randomized encoding.

Definition 6.2 (Semi-private randomized encoding (SPRE)) Let $g : \{0,1\}^n \to \{0,1\}$ be a boolean function. We say that a function $\hat{g} : \{0,1\}^n \times \{0,1\}^{m(n)} \to \{0,1\}^{s(n)}$ is a *semi-private randomized encoding* (SPRE) of g with error ε if the following conditions hold:

- **Statistical-correctness.** There exists a polynomial-time decoder B, such that for every $x \in \{0,1\}^n$ it holds that $\Pr[B(1^n, \hat{g}(x, U_{m(n)})) \neq g(x)] < \varepsilon(n)$.
- **One-sided privacy.** There exists a probabilistic polynomial-time simulator S_0, such that for every $x \in g^{-1}(0) \cap \{0,1\}^n$, it holds that $S_0(1^n) \equiv \hat{g}(x, U_{m(n)})$.
- **One-sided exposure.** There exists a polynomial-time exposure algorithm E_1, such that for every $x \in g^{-1}(1) \cap \{0,1\}^n$ $\Pr[E_1(1^n, \hat{g}(x, U_{m(n)})) \neq x] < \varepsilon(n)$.

By default, ε is a negligible function in n.

We turn to the question of constructing degree-2 SPRE. Since we will only need to encode functions that depend on a small number of inputs, it will be convenient to construct such an encoding based on the DNF representation of the function.

Construction 6.1 (SPRE for canonic DNF) *Let $g : \{0, 1\}^d \to \{0, 1\}$ be a boolean function. Let $\bigvee_{i=1}^{k} T_i$ be its unique canonic DNF representation. That is, for each $\alpha \in \{0, 1\}^d$ such that $g(\alpha) = 1$ there exists a corresponding term $T_i(x)$ which evaluates to 1 if and only if $x = \alpha$. We encode such T_i by the degree-2 function $\hat{T}_i(x, r) = \langle x - \alpha, r \rangle$, where $\langle \cdot, \cdot \rangle$ denotes inner product over $\mathbb{F}_2$. Let t be some integer (later used as a security parameter). Then, the degree-2 function $\hat{g}$ is defined by concatenating t copies of $\hat{T}_i$ (each copy with independent random inputs $r_{i,j}$) for each of the k terms. Namely,*

$$\hat{g}\big(x, (r_{i,j})_{i \in [k], j \in [t]}\big) \stackrel{\text{def}}{=} \big((\hat{T}_1(x, r_{1,j}))_{j=1}^{t}, \ldots, (\hat{T}_k(x, r_{k,j}))_{j=1}^{t}\big),$$

where $r_{i,j} \in \{0, 1\}^d$.

Lemma 6.3 *The encoding $\hat{g}$ (defined in Construction 6.1) is an SPRE of g with error $k \cdot 2^{-t}$. The time complexity of the simulator, decoder, and exposing algorithms is $\mathrm{poly}(tkd)$.*

Proof Let $\bigvee_{i=1}^{k} T_i$ be the canonic DNF representation of g. We view the variables $r_{i,j}$ of $\hat{T}$ as the random input of the encoding. Observe that if $T_i(x) = 1$ then $\hat{T}_i(x, r) = 0$ for every r. On the other hand, if $T_i(x) = 0$ then $\hat{T}_i(x, U_d)$ is distributed uniformly over $\mathbb{F}_2$ (since $\hat{T}_i$ is an inner product of a random vector with a non-zero vector). Therefore, when $g(x) = 0$, the output of all the copies of each $\hat{T}_i$ are distributed uniformly and independently over $\mathbb{F}_2$. Hence, we can perfectly simulate $\hat{g}(x; r)$ in time $O(tk)$.

If $g(x) = 1$ then there exists a single term T_i that equals to one and the other terms equal to zero (since this is a canonic DNF); thus all the copies of $\hat{T}_i$ equal to zero while the other $\hat{T}_j$'s are distributed uniformly and independently over $\mathbb{F}_2$. The extraction algorithm will locate the first i for which all the copies of $\hat{T}_i$ equal to zero, and output the unique x which satisfies the term T_i. This algorithm errs only if there exists another i for which $T_i(x)$ is not satisfied but all the copies of $\hat{T}_i$ equal to zero. This event happens with probability at most $k2^{-t}$.

Similarly, the decoder outputs 1 if and only if there exists an i for which all the copies of $\hat{T}_i$ equal to zero. Therefore the decoder never errs when $g(x) = 1$ and errs with probability at most $k2^{-t}$ when $g(x) = 0$. $\square$

Construction 6.1 yields an efficient degree-2 SPRE for functions whose canonic-DNF representation is efficiently computable. Specifically, we will be interested in functions $f : \{0, 1\}^n \to \{0, 1\}^l$ with logarithmic locality. In this case, we can efficiently encode each output with a degree-2 SPRE via the above construction.

6.3.2 Encoding a Function via SPRE

We move on to study the effect of applying an SPRE to a function. To this aim, we define a new operation on functions called *random exposure*.

Definition 6.3 (Random exposure of boolean function) Let $f : \{0, 1\}^n \to \{0, 1\}$ be a boolean function. The *random exposure* of f is the function $f_{\exp}(x, b)$ which maps $x \in \{0, 1\}^n$ and $b \in \{0, 1\}$ to the triple $(f(x), b, z)$ where

$$z = \begin{cases} x & \text{if } b = 1, \\ 0^n & \text{if } b = 0. \end{cases}$$

Observe that SPRE exposes the input when the output evaluates to one, while $f_{\exp}$ tosses a random coin b that determines whether to reveal the input or not. Still, it is not hard to move from 1-exposure to random exposure by padding the output with a random bit and revealing this bit.

Formally, fix some efficiently computable boolean function $f : \{0, 1\}^n \to \{0, 1\}$. Let $g(x, y) = f(x) \oplus y$ and let $\hat{g}(x, y, r)$ be an SPRE of g. We will show that the function $h(x, y, r) = (\hat{g}(x, y, r), y)$ statistically encodes $f_{\exp}$ under the generalized notion of randomized encoding.

Lemma 6.4 *Assume that $\hat{g}$ is an SPRE of g. Then, the function $h(x, y, r)$ is a (generalized) randomized encoding of the function $f_{\exp}(x, b)$.*

Proof Let $h'(x, b, r) = h(x, f(x) \oplus b, r)$. It is not hard to see that h is isomorphic to h' via the mapping $(x, b, r) \mapsto (x, f(x) \oplus b, r)$. Furthermore, this mapping is efficiently computable and efficiently invertible since f is efficiently computable. It is left to show that h' is a (standard) randomized encoding of the function $f_{\exp}(x, b)$ with statistical correctness and perfect privacy.

We begin with correctness. Given $h'(x, b, r) = (\hat{g}(x, y, r), y)$ for $y = f(x) \oplus b$ and a uniformly chosen r, we apply the decoder of $\hat{g}$ to the first entry and recover, with all but negligible probability, the value $g(x, y) = f(x) \oplus y = b$. By XOR-ing this with y we recover $f(x)$. Knowing both $f(x)$ and b, it is left to recover x in case $b = 1$. Indeed, when $b = 1$ the function $g(x, y) = b$ also evaluates to 1, and so we can apply the extraction algorithm to $\hat{g}(x, y, r)$ and recover x with all but negligible probability, as required. In both cases the efficiency of the decoder follows from the efficiency of the original decoder and extraction algorithm.

We move on to privacy. Given an output $(f(x), b, z)$ of $f_{\exp}$, we compute $y = f(x) \oplus b$ and sample $\hat{g}(x, y, r)$ as follows: (1) if $g(x, y) = f(x) \oplus y = 1$ (i.e., $b = 1$) the input x is available (via z) and so we simply compute $\hat{g}(x, y, r)$ for a uniformly chosen r; (2) if $g(x, y) = f(x) \oplus y = 0$ (i.e., $b = 0$), we sample $\hat{g}(x, y, r)$ via the perfect one-sided simulator S_0 of $\hat{g}$. In any case, the simulation is perfect and efficient. $\qquad\square$

We move on to the case of multi-output functions. In the following let $f :$ $\{0, 1\}^n \to \{0, 1\}^l$ be a function where f_i denotes the boolean function computing the i-th bit of f and $K(i)$ is the set of inputs that influence the i-th output. For a string x we let $x_{K(i)}$ denote the restriction of x to the indices in $K(i)$. The random exposure $f_{\exp}$ of a non-boolean function f is defined by concatenating the random exposures of each of the outputs. Formally,

Definition 6.4 (Random exposure of general function) The *random exposure* $f_{\exp}$ of a function $f : \{0, 1\}^n \to \{0, 1\}^l$ maps $x \in \{0, 1\}^n$ and $b_1, \ldots, b_l \in \{0, 1\}$ to $(f_{i,\exp}(x_{K(i)}, b_i))_{i \in [l]}$, where $f_{i,\exp}$ is the random exposure of the boolean function f_i.

By simple concatenation we extend Lemma 6.4 to the multi-output case.

Construction 6.2 (From SPRE to RE of $f_{\exp}$) *For a function* $f : \{0, 1\}^n \to \{0, 1\}^l$ *define the following functions for every* $i \in [l]$:

$$g_i(x_{K(i)}, y_i) \overset{\text{def}}{=} f_i(x_{K(i)}) \oplus y_i, \quad \text{where } f_i \text{ computes the } i\text{-th output of } f,$$

$$h_i(x_{K(i)}, y_i, r_i) \overset{\text{def}}{=} (\hat{g}_i(x_{K(i)}, y_i, r_i), y_i), \quad \text{where } \hat{g}_i \text{ is an SPRE of } g_i,$$

$$h(x, (y_i, r_i)_{i \in [l]}) \overset{\text{def}}{=} (h_i(x_{K(i)}, y_i, r_i))_{i \in [l]}.$$

Lemma 6.5 *The function* $h(x, y, r)$ *is a GRE of the function* $f_{\exp}(x, b)$.

Proof For every $i \in [l]$, let $h_i'(x_{K(i)}, b_i, r_i) = h_i(x_{K(i)}, f_i(x_{K(i)}) \oplus b_i, r_i)$. The proof of Lemma 6.4, shows that h_i' is a standard encoding of $f_{i,\exp}(x_{K(i)}, b_i)$ with perfect privacy and statistical correctness. By the concatenation lemma (Lemma 3.2) the function

$$h'(x, (b_i, r_i)_{i \in [l]}) = (h_i'(x_{K(i)}, b_i, r_i))_{i \in [l]},$$

is a standard encoding of $f_{\exp}(x, b)$. Finally, the isomorphism between h and h' is identical to the one described in the proof of Lemma 6.4. $\square$

We can now derive the main result of this section.

Theorem 6.1 *An efficiently computable function* $f : \{0, 1\}^n \to \{0, 1\}^{l(n)}$ *with output locality* $O(\log n)$ *can be encoded via degree-2 generalized encoding* h.

Proof Apply Construction 6.2 to f where each SPRE $\hat{g}_i$ is defined according to Construction 6.1. The theorem now follows from Lemmas 6.3 and 6.5. $\square$

6.4 Robust One-Way Function

In this section we define the notion of robust one-way functions (ROWF), we use the results of the previous section to argue that a local ROWF gives rise to a degree-2 one-way function, and, finally we present a construction of local ROWFs based on the one-wayness of Goldreich's function.

Definition 6.5 (**Robust one-way function**) Let $f : \{0, 1\}^n \rightarrow \{0, 1\}^{l(n)}$ be a polynomial-time computable function. We say that f is a *robust OWF* if its random exposure $f_{\exp}$ is a weak OWF.

Theorem 6.2 *If there exists a robust OWF with locality $O(\log n)$ then there exists a degree-2 one-way function with (optimal) output locality 3.*

Proof Let f be the ROWF and let $f_{\exp}$ be its random exposure which is, by definition, a weak one-way function. Since f has logarithmic locality we can apply Theorem 6.1 and get a degree-2 generalized encoding h of $f_{\exp}$. By Lemma 6.1, h is a distributional one-way function. Finally, by Lemma 6.2, a degree-2 distributional OWF can be transformed into a degree-2 OWF with output locality 3. $\square$

6.4.1 A Candidate Robust One-Way Function

Let m and n be integers corresponding to input length and output length. Let $M \in \{0, 1\}^{m \times n}$ be a binary matrix with d_i ones in the i-th row M_i. Through this section we abuse notation and view M_i both as a row vector and as the set $\{j | M_{i,j} = 1\}$. Correspondingly, we let $|M_i|$ denote the size d_i of the set M_i. For a real number $p \in (0, 1)$, denote by $\mathcal{M}_{m,n,p}$ the probability distribution over matrices $M \in \{0, 1\}^{m \times n}$ where each entry is chosen to be one with probability p independently of the other entries.

For a matrix M and a list of m predicates $P = (P_i)_{i=1}^m$ where $P_i : \{0, 1\}^{|M_i|} \rightarrow \{0, 1\}$, we define the function $f_{M,P} : \{0, 1\}^n \rightarrow \{0, 1\}^m$ via the mapping

$$x \mapsto \left(P_1(x_{M_1}), \ldots, P_m(x_{M_m}) \right).$$

Namely, the i-th output is computed by applying P_i to the string x restricted to the set M_i. For $m = m(n)$ and $p = p(n)$ we let $\mathcal{F}_{m,n,p}$ denote the collection of functions $f_{M,P}$ where M is chosen from $\mathcal{M}_{m,n,p}$ and the i-th predicate $P_i : \{0, 1\}^{|M_i|} \rightarrow \{0, 1\}$ is a random predicate whose truth table is sampled uniformly from $\{0, 1\}^{2^{|M_i|}}$. We assume that the collection $\mathcal{F}_{n,n,d/n}$ is slightly hard to invert for some constant d. Observe that in this case the expected output locality is d, and, by a Chernoff bound, with all but negligible probability, all outputs have locality at most $\log n$ and so P_i can be described by a polynomial-size string and the function can be evaluated in polynomial time.

Intractability Assumption 6.1 (**Random local function**) There exists a constant d and a polynomial $p(\cdot)$, such that for every (non-uniform) polynomial-time algorithm, A, and all sufficiently large n's

$$\Pr\left[A\left(1^n, M, P, f_{M,P}(x)\right) \notin f_{M,P}^{-1}\left(f_{M,P}(x)\right)\right] > \frac{1}{p(n)},$$

where $x \leftarrow U_n$, $M \leftarrow \mathcal{M}_{m,n,d/n}$ and $P = (P_i)_{i=1}^n$ and each $P_i : \{0,1\}^{|M_i|} \to \{0,1\}$ is chosen uniformly at random.

The collection $\mathcal{F}_{n,n,d/n}$ used in Assumption 6.1 is a variant of a candidate OWF proposed by Goldreich [69] (the main difference being that [69] uses the same predicate for all output bits). It is shown in [69] that if the matrix M satisfies some expansion property, then the function $f_{M,P}$ resists a natural class of inverting attacks. Since a random matrix has good expansion with high probability, our variant resists the same class of attacks.

Based on the one-wayness of $\mathcal{F}_{n,n,d/n}$ we show that the related collection $\mathcal{F}_{2n,e^d n,d/n}$ is robust one-way. Recall that a function $h \in \mathcal{F}_{2n,e^d n,d/n}$ is a random function with $e^d n$ inputs (here e denotes the base of the natural logarithm), $2n$ outputs and expected output locality of $d \cdot e^d$.

Theorem 6.3 *If Assumption 6.1 holds with respect to the constant d, then the collection $\mathcal{F}_{2n,e^d n,d/n}$ is robust one-way.*

Roughly speaking, we show that a random function $f_{M,P} \leftarrow \mathcal{F}_{n,n,d/n}$ can be embedded, with noticeable probability, in the function $h_{\exp}$, which is the random exposure of $h_{M',P'} \leftarrow \mathcal{F}_{2n,e^d n,d/n}$. Thus, $h_{\exp}$ is slightly hard to invert and $\mathcal{F}_{2n,e^d n,d/n}$ is robust one-way. To get some intuition, assume that exactly half of the outputs of $h_{M',P'} \leftarrow \mathcal{F}_{2n,e^d n,d/n}$ are exposed. Then there are exactly n exposed outputs and n unexposed outputs. In this case, an input node is not exposed (i.e., is *not* connected to an exposed output) with probability exactly $(1 - d/n)^n \approx e^{-d}$. Hence, the expected number of unexposed inputs is n, and so the matrix M' contains a non-exposed submatrix M with n inputs and n outputs. It turns out that the function $f_{M,P}$ can be embedded in this submatrix. The full proof of Theorem 6.3 is deferred to Sect. 6.4.2.

Observe that the locality of $h_{M,P}$ is bounded, with all but negligible probability, by $\log n$. (Indeed, by a multiplicative Chernoff bound the probability that an output will be connected to more than $\log n$ inputs is at most $e^{-\Omega(\log^2 n)}$.) Hence, if we modify the collection and reject functions whose locality is larger than $\log n$, we will get a robust one-way function with locality $\log n$. Also the truth tables P_i are of size at most n and are given explicitly in the description of $h_{M,P}$. Thus, we can apply Theorem 6.2 and construct a collection of degree-2 OWFs in $\mathbf{NC}_3^0$ based on Assumption 6.1.

Corollary 6.1 *Under Assumption 6.1, there exists a collection of degree-2 OWF in $\mathbf{NC}_3^0$.*

6.4.2 Proof of Theorem 6.3

The overall strategy is to embed an output of a function $f_{M,P} \leftarrow \mathscr{F}_{n,n,d/n}$ in an output of a random exposure $h_{\exp}$ of $h_{M',P'} \leftarrow \mathscr{F}_{2n,e^d n,d/n}$ and then show that an inverter $\hat{A}$ for $h_{\exp}$ can be used to invert $f_{M,P}$.

Formally, let $p(n)$ be the polynomial guaranteed by the assumption that $\mathscr{F}_{n,n,d/n}$ is weakly one-way, and assume, towards a contradiction, that the random exposure of $h \leftarrow \mathscr{F}_{2n,e^d n,d/n}$ is not weakly one-way. Specifically, let $\varepsilon(n)$ be a function which is smaller than $\frac{1}{n^2 p(n)}$ for infinitely many n's. Assume that there exists an efficient algorithm $\hat{A}$ that with probability $1 - \varepsilon(n)$ inverts a random output

$$\phi = (M_i, P_i, y_i, b_i, z_i)_{i=1}^{2n}$$

chosen from the *uniform distribution* Φ_n defined by letting: $M \leftarrow \mathscr{M}_{2n,e^d n,d/n}$, $P_i \leftarrow \{0,1\}^{2^{|M_i|}}$, $x \leftarrow \{0,1\}^{e^d n}$, $b = (b_1, \dots, b_{2n}) \leftarrow \{0,1\}^{2n}$, $y_i = P_i(x_{M_i})$ and $z_i = x_{M_i}$ if $b_i = 1$ and $0^{|M_i|}$ otherwise.

We would like to analyze the success probability of $\hat{A}$ on a restricted class of "typical" instances. Given a tuple ϕ, we say that an output i is *exposed* if $b_i = 1$ and say that an input j is *exposed* if it participates in an exposed output, i.e., if $M_{i,j} = 1$ for some exposed output i. We say that ϕ is *typical* if the number of exposed outputs and exposed inputs is exactly n and all the outputs are connected to at most $\log n$ inputs. We say that a typical tuple ϕ is in *canonical* form if the first n inputs and the first n outputs are exactly the ones which get exposed. Let T_n (respectively, C_n) be the uniform distribution Φ_n conditioned on selecting a typical (respectively, canonical) instance.

It turns out that the algorithm $\hat{A}$, which is guaranteed to invert random instances $\phi \leftarrow \Phi_n$ with probability $1 - \varepsilon$, inverts typical instances with probability $1 - \Omega(n\varepsilon)$. Furthermore, $\hat{A}$ can be easily modified into an inverter for canonical instances with similar success probability.

Lemma 6.6 *The inverter $\hat{A}$ inverts a typical instance $\phi \leftarrow T_n$ with probability at least $1 - \Omega(n\varepsilon)$. Furthermore, there exists an efficient algorithm B that inverts a random canonical instance $\phi \leftarrow C_n$ with probability at least $1 - \Omega(n\varepsilon)$.*

The first item is proven by showing that a random instance is typical with probability $\Omega(1/n)$, and the second item is proven by noting that a canonical instance can be randomized into a typical instance by randomly permuting the inputs and the outputs. See Sect. 6.4.2.1 for a full proof.

The next lemma, whose proof is deferred to Sect. 6.4.2.2, shows that a random output ψ of $\mathscr{F}_{n,n,d/n}$ can be embedded in a random canonical output of $\mathscr{F}_{2n,e^d n,d/n}$.

Lemma 6.7 *There is an efficient transformation α which takes a random output ψ of $\mathscr{F}_{n,n,d/n}$, aborts with failure with negligible probability, and otherwise (conditioned on not aborting) outputs a random canonical output $\phi \leftarrow C_n$ such that if $x = (x_1, \dots, x_{2n})$ is a preimage of ψ then the prefix $(x_1, \dots, x_n)$ is a preimage of ϕ.*

We can now complete the proof of Theorem 6.3. By Lemma 6.6, the existence of the inverter $\hat{A}$ which inverts Φ_n with advantage $1 - \varepsilon$ implies the existence of an efficient algorithm B that inverts random canonical instances $\phi \leftarrow C_n$ with probability $1 - \Omega(1/(n^2 p(n)))$. Hence, by Lemma 6.7, B can be used to invert $\mathscr{F}_{n,n,d/n}$ with similar probability as follows. Map an output of $\mathscr{F}_{n,n,d/n}$ to an output $\phi \leftarrow C_n$ via α, use B to find a preimage x' and output the first n bits of x'. The success probability is $1 - \Omega(1/(n^2 p(n)))$, in contradiction to Assumption 6.1.

6.4.2.1 Proof of Lemma 6.6

We will need the following claim:

Claim 6.1 *For some constant $a > 0$, $\Pr_{\phi \leftarrow \Phi_n}[\phi \text{ is typical}] > \frac{a}{n}$.*

Proof Since each of the $2n$ output vertices is exposed with probability $\frac{1}{2}$, the probability that exactly n outputs are exposed is $\binom{2n}{n} \cdot 2^{-2n} = \Omega(1/\sqrt{n})$. Conditioned on this event, we claim that the probability that there are exactly n exposed inputs is $\Omega(1/\sqrt{n})$. Indeed, an input is exposed with probability $(1 - d/n)^n$ since each edge exists with probability d/n, and therefore the number of exposed inputs is distributed according to the binomial distribution. Therefore it suffices to show that $b(n; n \cdot e^d, (1 - d/n)^n) > \Omega(1/\sqrt{n})$, where $b(k; m, p)$ is the probability to have exactly k successes out of a sequence of m independent Bernoulli trials each with a probability p of success. Indeed, letting $p = (1 - d/n)^n$ and $m = n \cdot e^d$, we can write:

$$b(n; m, p) = b(\lceil pm \rceil + O(1); m, p) > \Omega(b(\lceil pm \rceil; m, p)) > \Omega(1/\sqrt{n}),$$

where the first equality follows from the estimate $(1 - d/n)^n = e^{-d} - O(1/n)$ (derived from the Taylor expansion), the second inequality follows from the smoothness of $b(k; m, p)$ with respect to k, namely, the ratio $b(k; m, p)/b(k + 1; m, p) = (1 - p)k/(m - k)p$ is constant when $k = \Theta(m)$ and $p \in (0, 1)$ is a constant, and the third inequality follows from the standard estimation $b(\lceil mp \rceil; m, p) = \Omega(1/\sqrt{n})$ where $p \in (0, 1)$ is bounded away from 0 and 1. Finally, by a Chernoff bound, the probability that some output is connected to more than $\log n$ inputs is negligible $(ne^{-\Omega(\log^2 n)})$, and the claim follows. $\square$

We can now prove Lemma 6.6. The first item follows from Bayes' law as $\Pr[\hat{A} \text{ inverts } T_n]$ equals to

$$\frac{\Pr[\hat{A} \text{ inverts } \Phi_n] - \Pr[\Phi_n \text{ is not typical}] \cdot \Pr[\hat{A} \text{ inverts } \Phi_n | \Phi_n \text{ is not typical}]}{\Pr[\Phi \text{ is typical}]}.$$

By applying Claim 6.1 and bounding $\Pr[\hat{A} \text{ inverts } \Phi_n | \Phi_n \text{ is not typical}]$ by 1, we get:

$$\Pr[\hat{A} \text{ inverts } T_n] > (1 - \varepsilon - (1 - a/n))n/a > 1 - n\varepsilon/a,$$

where a is a positive constant.

To prove the second item we note that one can easily map T_n to C_n by randomly permuting the order of the inputs and the order of the outputs. Formally, given a canonical instance $\phi = (M_i, P_i, y_i, b_i, z_i)_{i=1}^{2n}$ we define the inverter B as follows:

- Choose a random output permutation $\sigma : [2n] \to [2n]$ and let $\phi' = (M_i', P_i', y_i', b_i', z_i')_{i=1}^{2n}$ where

$$\left(M_{\sigma(i)}', P_{\sigma(i)}', y_{\sigma(i)}', b_{\sigma(i)}', z_{\sigma(i)}'\right) = (M_i, P_i, y_i, b_i, z_i).$$

- Choose a random input permutation $\pi : [ne^d] \to [ne^d]$ and define

$$\phi'' = \left(M_i'', P_i'', y_i', b_i', z_i''\right)_{i=1}^{2n}$$

as follows. (1) $M_i'' = \{\pi(j) | j \in M_i\}$. (2) Let $\pi_i : [|M_i|] \to [|M_i|]$ be the permutation that maps j to k if the j-th largest element in M_i' is mapped by π to be the k-th largest element in M_i''. Then, the predicate P_i'' is defined by permuting the input variables of P_i' under π_i. (3) The string z_i'' is defined by permuting the entries of z_i' under π_i.

- Finally, invoke $\hat{A}$ on ϕ'', copy the output to x'', permute the entries of x'' under π^{-1}, and output the resulting string x.

It is not hard to see that a random canonical tuple ϕ is mapped by B into a random typical tuple ϕ''. Furthermore, if x'' is a preimage of ϕ'', then x is a preimage of ϕ. Hence, B succeeds with probability $1 - \Omega(n\varepsilon)$.

6.4.2.2 Proof of Lemma 6.7

In the following we let $\mathscr{L}_{a,b,p}$ denote the distribution $\mathscr{M}_{a,b,p}$ conditioned on the event that none of the columns is an all-zero column, and none of the rows have Hamming weight larger than $\log n$. For our setting of parameters (i.e., $a = \Theta(n), b = \Theta(n)$ and $p = \Theta(1/n)$), such a distribution can be sampled efficiently with negligible failure probability.[2]

Given a tuple $\psi = (M_i, P_i, y_i)_{i=1}^{n}$ we compute $\phi = (M_i', P_i', y_i', b_i, z_i)_{i=1}^{2n}$ as follows:

1. **Padding the matrix.** Sample a $2n \times (e^d n)$ matrix

$$M' \leftarrow \begin{pmatrix} M & \mathscr{M}_{n,ne^d-n,d/n} \\ \mathbf{0}_{n\times n} & \mathscr{L}_{n,ne^d-n,d/n} \end{pmatrix}.$$

Halt with failure if one of the rows of M' has Hamming weight larger than $\log n$.

---

[2]Sample a matrix from $\mathscr{M}_{a,b,p}$ conditioned on the event that none of the columns is an all-zero column (a task which can be achieved efficiently by rejecting and resampling zero columns), and fail it if the resulting matrix has a row of weight larger than $\log n$. To see that rejection happens with negligible probability, observe that for a fixed row, each column contributes 1 with probability $p/(1-p)^a = \Theta(1/n)$, independently of the other columns, and therefore, by a Chernoff bound, the weight exceeds $\log n$ with negligible probability.

2. **Exposed inputs**. Choose a random $ne^d - n$ string $x_{n+1} \ldots x_{ne^d}$.
3. **Unexposed outputs**. For $i \in \{1, \ldots, n\}$ let

$$y_i' = y_i, \qquad b_i = 0, \qquad z_i = 0^{|M_i'|}.$$

Choose $P_i' : \{0, 1\}^{|M_i'|} \to \{0, 1\}$ uniformly at random subject to the constraint

$$P_i'(\beta, x_{M_i' \setminus M_i}) = P_i(\beta) \quad \text{for every } \beta \in \{0, 1\}^{|M_i|}.$$

Namely, the restricted predicate $P_i'(\cdot, x_{M_i' \setminus M_i})$ is equal to P_i.
4. **Exposed outputs**. For each $i \in \{n + 1, \ldots, 2n\}$ let

$$P_i' \leftarrow \{0, 1\}^{2^{|M_i'|}}, \qquad y_i' = P_i'(x_{M_i'}), \qquad b_i = 1, \qquad z_i = x_{M_i'}$$

(Since M_i' contains only "exposed inputs" ($n + 1 \le j \le ne^d$) we can compute y_i' and z_i.)

Assume that ψ is uniformly distributed, namely $M \leftarrow \mathcal{M}_{n,n,d/n}$ and for $i \in [n]$, $P_i \leftarrow \{0, 1\}^{2^{|M_i'|}}$ and $y_i = P_i(x)$ where $x \leftarrow \{0, 1\}^n$. Then, by the definition of $\mathcal{L}$ and by a multiplicative Chernoff bound, the failure probability in the first step is negligible. We further claim that, conditioned on not failing, ϕ is distributed according to C_n. Indeed, $b = 1^n 0^n$, the first n rows of M' are distributed according to $\mathcal{M}_{n,e^d n,d/n}$ and the last rows according to $0_{n \times n} \mathcal{L}_{n,ne^d-n,d/n}$. Furthermore, since the P_i's are random so are the P_i''s, and since $y_i = P_i(x_{M_i})$ for $x \leftarrow \{0, 1\}^n$, we have, by construction, that $y_i' = P_i'(x_{M_i'})$ where $x \leftarrow \{0, 1\}^{2n}$. Finally, again, by construction, for $n + 1 \le i \le 2n$ we have that $z_i = x_{M_i'}$ and $z_i = 0^{|M_i'|}$ otherwise.

Now suppose that $x' = (x_1', \ldots, x_{2n}')$ is a preimage of ψ. We show that, in this case, the prefix $(x_1', \ldots, x_n')$ is a preimage of ϕ. Since all the inputs $n + 1 \le i \le 2n$ are exposed, we have that $x_{n+1,\ldots,2n}' = x_{n+1,\ldots,2n}$. In addition, for $1 \le i \le n$ we have

$$y_i = P_i'\big(x_{M_i}', x_{M_{i,n+1\ldots2n}'}'\big) = P_i'\big(x_{M_i}', x_{M_{i,n+1\ldots2n}'}\big) = P_i\big(x_{M_i}'\big),$$

where the last equality follows from the definition of P_i'. Therefore, $f_{M,P}(x_1', \ldots, x_n') = y$, as claimed. This completes the proof of Lemma 6.7.

6.5 Addendum: Notes and Open Questions

Goldreich's Function   Assumption 6.1 asserts that the collection of random local functions $\mathcal{F}_{n,n,d/n}$ is one-way. As already mentioned this assumption originated in the work of Goldreich [69]. In the last few years, the one-wayness of this collection (or close variants of it) was widely studied.[3] In [6, 47, 69] it is shown that a large class of algorithms (including ones that capture DPLL-based heuristics) fail to invert $\mathscr{F}$ in polynomial time. These results are further supported by the experimental study of [47, 120] which employs, among other attacks, SAT-solvers. In addition, a self-amplification theorem was proved in [39] showing that when the output length is sufficiently large, i.e., $n + \Omega_d(n)$, if $\mathscr{F}$ is hard-to-invert over tiny subexponentially small fraction of the inputs with respect to subexponential-time algorithms, then the same ensemble is actually hard-to-invert over almost all inputs (with respect to subexponential-time algorithms).

The results of this chapter (Theorem 6.3) essentially show that, for proper parameters, the collection $\mathscr{F}$ achieves some form of *leakage resilience* under random exposure. This is contrasted with the results of Bogdanov and Qiao [38] who show that when a *single predicate P* is being used for all outputs and when the output length is sufficiently long $m = n + \Omega_d(n)$, the collection $\mathscr{F}$ can be fully inverted in the presence of leakage.

Standard RE with Degree 2 In Sect. 6.3 it is shown that non-trivial functions can be encoded by degree-2 *semi-private* randomized encoding. This leaves open the possibility of achieving standard degree-2 encoding with statistical privacy and statistical correctness. A positive result would lead to round-optimal secure computation protocols in the statistical setting of [27].

Locality 2 Over Larger Alphabet As already mentioned, the task of inverting a 2-local binary function reduces to solving an instance of 2-SAT and can therefore be implemented in polynomial time. However, this attack does not generalize to 2-local functions over a larger *non-binary* alphabet. Indeed, it is shown in [55] that, assuming the existence of one-way functions in log-space (or more generally in **CREN**), there are one-way functions with locality 2 over a (large) constant-size alphabet.

---

[3] Another line of work studied the pseudorandomness of the collection. See Sect. 7.6.

Chapter 7
On Pseudorandom Generators with Linear Stretch in $\mathbf{NC}^0$

Abstract We consider the question of constructing cryptographic pseudorandom generators in $\mathbf{NC}^0$ with large stretch. Our previous constructions of such PRGs were limited to stretching a seed of n bits to $n + o(n)$ bits. This leaves open the existence of a PRG with a linear (let alone superlinear) stretch in $\mathbf{NC}^0$. In this chapter we study this question and obtain the following main results: (1) We show that the existence of a linear-stretch PRG in $\mathbf{NC}^0$ implies non-trivial hardness of approximation results *without relying on PCP machinery*. In particular, it implies that Max3SAT is hard to approximate to within some multiplicative constant. (2) We construct a linear-stretch PRG in $\mathbf{NC}^0$ under a specific intractability assumption related to the hardness of decoding "sparsely generated" linear codes. Such an assumption was previously conjectured by Alekhnovich (Proc. of 44th FOCS, pp. 298–307, 2003).

7.1 Introduction

A cryptographic pseudorandom generator (PRG) (cf. Definition 4.3) is a deterministic function that stretches a short random seed into a longer string that cannot be distinguished from random by any polynomial-time observer. In this chapter, we study the existence of PRGs that both (1) admit fast parallel computation and (2) stretch their seed by a significant amount.

Considering the first goal alone, we showed in Chap. 4 that the ultimate level of parallelism can be achieved under most standard cryptographic assumptions. Specifically, any log-space computable OWF (the existence of which follows, for example, from the intractability of factoring) can be efficiently "compiled" into a PRG in $\mathbf{NC}^0$. However, the PRGs produced by this compiler only stretch their seed by a sublinear amount: from n bits to $n + O(n^\varepsilon)$ bits for some constant $\varepsilon < 1$. Thus, these PRGs do not meet our second goal.

Considering the second goal alone, even a PRG that stretches its seed by just one bit can be used to construct a PRG that stretches its seed by any polynomial number of bits [70, Sect. 3.3.2]. However, all known constructions of this type are inherently sequential. Thus, we cannot use known techniques for turning an $\mathbf{NC}^0$ PRG with a sublinear stretch into one with a linear, let alone superlinear, stretch.

The above state of affairs leaves open the existence of a *linear-stretch* PRG (LPRG) in $\mathbf{NC}^0$; namely, one that stretches a seed of n bits into $n + \Omega(n)$ out-

B. Applebaum, *Cryptography in Constant Parallel Time*,
Information Security and Cryptography,
DOI 10.1007/978-3-642-17367-7_7, © Springer-Verlag Berlin Heidelberg 2014

put bits.[1] (In fact, there was no previous evidence for the existence of LPRGs even in the higher complexity class $\mathbf{AC}^0$.) This question is the main focus of this chapter. The question has a very natural motivation from a cryptographic point of view. Indeed, most cryptographic applications of PRGs either require a linear stretch (for example Naor's bit commitment scheme [114][2]), or alternatively depend on a larger stretch for efficiency (this is the case for the standard construction of a stream cipher or stateful symmetric encryption from a PRG, see [72]). Thus, the existence of an LPRG in $\mathbf{NC}^0$ would imply better parallel implementations of other cryptographic primitives.

7.1.1 Our Contribution

LPRG in $\mathbf{NC}^0$ Implies Hardness of Approximation We give a very different, and somewhat unexpected, motivation for the foregoing question. We observe that the existence of an LPRG in $\mathbf{NC}^0$ *directly* implies non-trivial and useful hardness of approximation results. Specifically, we show (via a simple argument) that an LPRG in $\mathbf{NC}^0$ implies that Max3SAT (and hence all **MaxSNP** problems such as Max-Cut, Max2SAT and Vertex Cover [121]) cannot be efficiently approximated to within some multiplicative constant. This continues a recent line of work, initiated by Feige [58] and followed by Alekhnovich [5], that provides simpler alternatives to the traditional PCP-based approach by relying on stronger assumptions. Unlike these previous works, which rely on very specific assumptions, our assumption is of a more general flavor and may serve to further motivate the study of cryptography in $\mathbf{NC}^0$. On the downside, the conclusions we get are quantitatively weaker than the ones implied by the PCP theorem. In contrast, some inapproximability results from [5, 58] could not be obtained using PCP machinery. It is instructive to note that by applying our general argument to the sublinear-stretch PRGs in $\mathbf{NC}^0$ from Chap. 4 we only get "uninteresting" inapproximability results that follow from standard padding arguments (assuming $\mathbf{P} \neq \mathbf{NP}$). Furthermore, we do not know how to obtain stronger inapproximability results based on a superlinear-stretch PRG in $\mathbf{NC}^0$. Thus, our main question of constructing LPRGs in $\mathbf{NC}^0$ captures precisely what is needed for this application.

Constructing an LPRG in $\mathbf{NC}^0$ We present a construction of an LPRG in $\mathbf{NC}^0$ under a specific intractability assumption related to the hardness of decoding

---

[1]Recall that an $\mathbf{NC}^0$ LPRG can be composed with itself a constant number of times to yield an $\mathbf{NC}^0$ PRG with an arbitrary linear stretch. See Remark 4.3.

[2]In Chap. 5 we showed that there is an $\mathbf{NC}^0$ construction of a commitment scheme from an arbitrary PRG including one with sublinear stretch (see Corollary 5.2). However, this construction makes a non-black-box use of the underlying PRG, and is thus quite inefficient. The only known parallel construction that makes a black-box use of the PRG is Naor's original construction, which requires the PRG to have linear stretch.

"sparsely generated" linear codes. Such an assumption was previously made by Alekhnovich in [5]. The starting point of our construction is a modified version of a PRG from [5] that has a large output locality (that is, each output bit depends on many input bits) but has a simple structure. We note that the output distribution of this generator can be sampled in $\mathbf{NC}^0$; however the seed length of this $\mathbf{NC}^0$ sampling procedure is too large to gain any stretch. To solve this problem we observe that the seed has large entropy even when the output of the generator is given. Hence, we can regain the stretch by employing a randomness extractor in $\mathbf{NC}^0$ that uses a "sufficiently short" seed to extract randomness from sources with a "sufficiently high" entropy. We construct the latter by combining the known construction of randomness extractors from ε-biased generators [28, 113] with previous constructions of ε-biased generator in $\mathbf{NC}^0$ [112]. Our LPRG can be implemented with locality 4; the stretch of this LPRG is essentially optimal, as it is known that no PRG with locality 4 can have a *superlinear* stretch [112]. However, the existence of superlinear-stretch PRG with possibly higher (but constant) locality remains open.

By combining the two main results described above, one gets non-trivial inapproximability results under the intractability assumption from [5]. These (and stronger) results were *directly* obtained in [5] from the same assumption *without* constructing an LPRG in $\mathbf{NC}^0$. Our hope is that future work will yield constructions of LPRGs in $\mathbf{NC}^0$ under different, perhaps more standard, assumptions, and that the implications to hardness of approximation will be strengthened.

LPRG in $\mathbf{NC}^0$ and Expanders Finally, we observe that the input-output graph of any LPRG in $\mathbf{NC}^0$ enjoys some non-trivial expansion property. This connection implies that a (deterministic) construction of an LPRG in $\mathbf{NC}^0$ must use some non-trivial combinatorial objects. (In particular, one cannot hope that "simple" transformations, such as those given in Chap. 4, will yield LPRGs in $\mathbf{NC}^0$.) The connection with expanders also allows us to rule out the existence of *exponentially*-strong PRGs with *superlinear* stretch in $\mathbf{NC}^0$.

7.1.2 Related Work

The first application of average-case complexity to inapproximability was suggested by Feige [58], who derived new inapproximability results under the assumption that refuting 3SAT is hard on average on some natural distribution. In [5] Alekhnovich continued this line of research. He considered the problem of determining the maximal number of satisfiable equations in a linear system chosen at random, and made several conjectures regarding the average case hardness of this problem. He showed that these conjectures imply Feige's assumption as well as several new inapproximability results. While the works of Feige and Alekhnovich derived *new* inapproximability results (that were not known to hold under the assumption that $\mathbf{P} \neq \mathbf{NP}$), they did not rely on the relation with a standard cryptographic assumption or primitive, but rather used specific average case hardness assumptions tailored to their inapproximability applications. A relation between the security of a cryptographic primitive and approximation was implicitly used in [112], where an approximation algorithm for Max2LIN was used to derive an upper bound on the stretch of a PRG whose locality is 4.

Organization　The rest of this chapter is structured as follows. We begin with a discussion of notation and preliminaries (Sect. 7.2). In Sect. 7.3 we prove that an LPRG in $\mathbf{NC}^0$ implies that Max3SAT cannot be efficiently approximated to within some multiplicative constant. In Sect. 7.4 we present a construction of an LPRG in $\mathbf{NC}^0$. This construction uses an $\mathbf{NC}^0$ implementation of an ε-biased generator as an ingredient. A uniform construction of such an ε-biased generator is described in Sect. 7.4.4. Finally, in Sect. 7.5, we discuss the connection between LPRG in $\mathbf{NC}^0$ and expander graphs.

7.2 Preliminaries

Pseudorandom Generators　Recall that a PRG, $G : \{0, 1\}^n \to \{0, 1\}^{s(n)}$ is a deterministic function that stretches a short random seed into a longer pseudorandom string (see Definition 4.3). When $s(n) = n + \Omega(n)$ we say that G is a *linear-stretch* pseudorandom generator (LPRG). By default, we require G to be polynomial-time computable. It will sometimes be convenient to define a PRG by an infinite family of functions $\{G_n : \{0, 1\}^{m(n)} \to \{0, 1\}^{s(n)}\}_{n \in \mathbb{N}}$, where $m(\cdot)$ and $s(\cdot)$ are polynomials. Such a family can be transformed into a single function that satisfies Definition 4.3 via padding (see Remark 3.1). We will also rely on ε-*biased generators*, defined similarly to PRGs except that the pseudorandomness holds only against linear functions over $\mathbb{F}_2$. (See Definition 4.4.)

Expanders　In the following think of m as larger than n. We say that a bipartite graph $G = ((L = [m], R = [n]), E)$ is (K, α) expanding if every set of left vertices S of size smaller than K has at least $\alpha \cdot |S|$ right neighbors. A family of bipartite graphs $\{G_n\}_{n \in \mathbb{N}}$ where $G_n = ((L = [m(n)], R = [n]), E)$ is expanding if for some constants α and β and sufficiently large n the graph G_n is $(\beta \cdot m(n), \alpha)$ expanding. A family of $m(n) \times n$ binary matrices $\{M_n\}_{n \in \mathbb{N}}$ is expanding if the family of bipartite graphs $\{G_n\}_{n \in \mathbb{N}}$ represented by $\{M_n\}_{n \in \mathbb{N}}$ (i.e., M_n is the adjacency matrix of G_n) is expanding.

7.2.1 Some Useful Facts

We will use the following well-known bound on the sum of binomial coefficients.

Fact 7.1　*For $0 < p \le 1/2$ we have $\sum_{i=0}^{pn} \binom{n}{i} \le 2^{n H_2(p)}$.*

The *bias* of a Bernoulli random variable X is defined to be $|\Pr[X = 1] - \frac{1}{2}|$. We will need the following fact which estimates the bias of the sum of independent random coins (cf. [112, 131]).

Fact 7.2 *Let $X_1, \ldots, X_t$ be independent binary random variables. Suppose that for some $0 < \delta < \frac{1}{2}$ and every i it holds that $\mathrm{bias}(X_i) \leq \delta$. Then, $\mathrm{bias}(\bigoplus_{i=1}^{t} X_i) \leq \frac{1}{2}(2\delta)^t$.*

7.3 LPRG in NC^0 Implies Hardness of Approximation

In the following we show that if there exists an LPRG in NC^0 then there is no polynomial-time approximation scheme (PTAS) for Max3SAT; that is, Max3SAT cannot be efficiently approximated within some multiplicative constant $r > 1$. Recall that in the Max3SAT problem we are given a 3CNF boolean formula with s clauses over n variables, and the goal is to find an assignment that satisfies the largest possible number of clauses. The Max ℓ-CSP problem is a generalization of Max3SAT in which instead of s clauses we get s boolean constraints $C = \{C_1, \ldots, C_s\}$ of arity ℓ. Again, our goal is to find an assignment that satisfies the largest possible number of constraints. (Recall that a constraint C of arity ℓ over n variables is an ℓ-local boolean function $f : \{0, 1\}^n \to \{0, 1\}$, and it is satisfied by an assignment $(\sigma_1, \ldots, \sigma_n)$ if $f(\sigma_1, \ldots, \sigma_n) = 1$.)

A simple and useful corollary of the PCP Theorem [21, 22] is the inapproximability of Max3SAT.

Theorem 7.1 *Assume that $P \neq NP$. Then, there is an $\varepsilon > 0$ such that there is no $(1 + \varepsilon)$-approximation algorithm for Max3SAT.*

We will prove a similar result under the (stronger) assumption that there exists an LPRG in NC^0. Our proof, however, does not rely on the PCP Theorem.

Theorem 7.2 *Assume that there exists an LPRG in NC^0. Then, there is an $\varepsilon > 0$ such that there is no $(1 + \varepsilon)$-approximation algorithm for Max3SAT.*

The proof of Theorem 7.2 follows by combining the following Fact 7.3 and Lemma 7.1. The first fact shows that in order to prove that Max3SAT is hard to approximate, it suffices to prove that Max ℓ-CSP is hard to approximate. This standard result follows by applying Cook's reduction to transform every constraint into a 3CNF.

Fact 7.3 *Assume that, for some constants $\ell \in \mathbb{N}$ and $\varepsilon > 0$, there is no polynomial-time $(1 + \varepsilon)$-approximation algorithm for Max ℓ-CSP. Then there is an $\varepsilon' > 0$ such that there is no polynomial-time $(1 + \varepsilon')$-approximation algorithm for Max3SAT.*

Thus, the heart of the proof of Theorem 7.2 is showing that the existence of an LPRG in $\mathbf{NC}^0_\ell$ implies that there is no PTAS for Max ℓ-CSP.

Lemma 7.1 *Let ℓ be a positive integer, and $c > 1$ be a constant such that $G : \{0, 1\}^n \to \{0, 1\}^{cn}$ is an LPRG which is computable in $\mathbf{NC}^0_\ell$. Then, there is no $1/(1 - \varepsilon)$-approximation algorithm for Max ℓ-CSP, where $0 < \varepsilon < 1/2$ is a constant that satisfies $H_2(\varepsilon) < 1 - 1/c$.*

For $\varepsilon = 1/10$ (i.e., ≈ 1.1-approximation) the constant $c = 2$ will do, whereas for $\varepsilon = 0.49$ (i.e., ≈ 2-approximation) $c = 3500$ will do.

Proof Let $s = s(n) = cn$. Assume towards a contradiction that there exists an $1/(1 - \varepsilon)$-approximation algorithm for Max ℓ-CSP where $H_2(\varepsilon) < 1 - 1/c$. Then, there exists a polynomial-time algorithm A that given an ℓ-CSP instance ϕ outputs 1 if ϕ is satisfiable, and 0 if ϕ is ε-unsatisfiable (i.e., if every assignment fails to satisfy at least a fraction ε of the constraints). We show that, given such A, we can "break" the LPRG G; that is, we can construct an efficient (non-uniform) adversary that distinguishes between $G(U_n)$ and U_s. Our adversary B_n will (deterministically) translate a string $y \in \{0, 1\}^s$ into an ℓ-CSP instance ϕ_y with s constraints such that the following holds:

1. If $y \leftarrow G(U_n)$ then ϕ_y is always satisfiable.
2. If $y \leftarrow U_s$ then, with probability $1 - \text{neg}(n)$ over the choice of y, no assignment satisfies more than $(1 - \varepsilon)s$ constraints of ϕ_y.

Then, B_n will run A on ϕ_y and will output $A(\phi_y)$. The distinguishing advantage of B is $1 - \text{neg}(n)$ in contradiction to the pseudorandomness of G.

It is left to show how to translate $y \in \{0, 1\}^s$ into an ℓ-CSP instance ϕ_y. We use n boolean variables $x_1, \ldots, x_n$ that represent the bits of a hypothetical preimage of y under G. For every $1 \le i \le s$ we add a constraint $G_i(x) = y_i$ where G_i is the function that computes the i-th output bit of G. Since G_i is an ℓ-local function the arity of the constraint is at most ℓ.

Suppose first that $y \leftarrow G(U_n)$. Then, there exists a string $\sigma \in \{0, 1\}^n$ such that $G(\sigma) = y$ and hence ϕ_y is satisfiable. We move on to the case where $y \leftarrow U_s$. Here, we rely on the fact that such a random y is very likely to be far from every element in the range of G. More formally, define a set $\text{BAD}_n \subseteq \{0, 1\}^s$ such that $y \in \text{BAD}_n$ if ϕ_y is $(1 - \varepsilon)$-satisfiable; that is, if there exists an assignment $\sigma \in \{0, 1\}^n$ that satisfies at least a $(1 - \varepsilon)$ fraction of the constraints of ϕ_y. In other words, the Hamming distance between y and $G(\sigma)$ is at most εs. Hence, all the elements of BAD_n are εs-close (in Hamming distance) to some string in $\text{Im}(G)$. Therefore, the size of BAD_n is bounded by

$$|\text{Im}(G)| \cdot \sum_{i=0}^{\varepsilon s} \binom{s}{i} \le 2^n 2^{H_2(\varepsilon)s} = 2^{(1+cH_2(\varepsilon))n},$$

where the first inequality is due to Fact 7.1. Let $\alpha \overset{\text{def}}{=} c - (1 + c \cdot H_2(\varepsilon))$ which is a positive constant since $H_2(\varepsilon) < 1 - 1/c$. Hence, we have

$$\Pr_{y \leftarrow U_s} \left[\phi_y \text{ is } (1 - \varepsilon) \text{ satisfiable} \right] = |\text{BAD}_n| \cdot 2^{-s} \leq 2^{(1+cH_2(\varepsilon))n-cn} = 2^{-\alpha n} = \text{neg}(n),$$

which completes the proof.   □

Remark 7.1 Lemma 7.1 can tolerate some relaxations to the notion of LPRG. In particular, since the advantage of B_n is exponentially close to 1, we can consider an LPRG that satisfies a weaker notion of pseudorandomness in which the distinguisher's advantage is bounded by $1 - 1/p(n)$ for some polynomial $p(n)$.

Lemma 7.1 implies the following corollary.

Corollary 7.1 *Suppose there exists a PRG in $\mathbf{NC}_\ell^0$ with an arbitrary linear stretch; i.e., for every $c > 0$ there exists a PRG $G : \{0, 1\}^n \rightarrow \{0, 1\}^{c \cdot n} \in \mathbf{NC}_\ell^0$. Then, Max ℓ-CSP cannot be approximated to within any constant $\delta < 2$ that is arbitrarily close to 2.*

Remark 7.2 Corollary 7.1 is tight, as any CSP problem of the form $G(x) = y$ (for any $y \in \{0, 1\}^s$) can be easily approximated within a factor of 2. To see this, note that the function $G_i(x)$ which computes the i-th output bit of G must be balanced, i.e., $\Pr_x[G_i(x) = 1] = 1/2$. (Otherwise, since $G_i \in \mathbf{NC}^0$, the function G_i has a constant bias and so $G(U_n)$ cannot be pseudorandom.) Therefore, a random assignment is expected to satisfy $1/2$ of the constraints of the instance $G(x) = y$. This algorithm can be derandomized by using the method of conditional expectations.

Papadimitriou and Yannakakis [121] defined the complexity class **MaxSNP**, in which Max3SAT is complete in the sense that any problem in **MaxSNP** has a PTAS if and only if Max3SAT has a PTAS. Hence, we get the following corollary (again, without the PCP machinery).

Corollary 7.2 *Assume that there exists an LPRG in $\mathbf{NC}^0$. Then, all Max SNP problems (e.g., Max-Cut, Max2SAT, Vertex Cover) do not have a PTAS.*

7.4 A Construction of LPRG in $\mathbf{NC}^0$

7.4.1 Overview

We start with an informal description of our construction. Consider the following distribution: *fix* a sparse matrix $M \in \{0, 1\}^{m \times n}$ in which every row contains a constant number of ones, multiply it with a random n-bit vector x, and add a noise vector $e \in \{0, 1\}^m$ which is uniformly distributed over all m-bits vectors whose Hamming weight is $\lceil \mu \cdot m \rceil$. (For concreteness, think of $m = 5n$ and $\mu = 0.1$.) That is, we consider the distribution $d_\mu(M) \overset{\text{def}}{=} M \cdot x + e$, where all arithmetic is over $\mathbb{F}_2$.

Consider the distribution $d_{\mu+m^{-1}}(M)$ which is similar to the previous distribution except that this time the noise vector is uniformly distributed over m-bit vectors whose weight is $(\mu + 1/m) \cdot m = \mu m + 1$. Alekhnovich conjectured in [5, Conjecture 1] that for a proper choice of M these distributions are computationally indistinguishable. He also showed that if indeed this is the case, then $d_\mu(M)$ is pseudorandom; that is, $d_\mu(M)$ is computationally indistinguishable from U_m. Since the distribution $d_\mu(M)$ can be sampled (efficiently) by using roughly $n + \log \binom{m}{\mu \cdot m} \leq n + m H_2(\mu)$ random bits, it gives rise to a pseudorandom generator with linear stretch (when the parameters are chosen properly).

We would like to sample $d_\mu(M)$ by an **NC**^{0} function. Indeed, since the rows of M contains only a constant number of ones, we can easily compute the product Mx in **NC**^{0} (recall that M itself is fixed). Unfortunately, we do not know how to sample the noise vector e by an **NC**^{0} function.[3] To solve this, we change the noise distribution. That is, we consider a slightly different distribution $D_\mu(M)$ in which each entry of the noise vector e is chosen to be 1 with probability μ (independently of other entries). We adopt Alekhnovich's conjecture to this setting; namely, we assume that $D_\mu(M)$ cannot be distinguished efficiently from $D_{\mu+m^{-1}}(M)$. (In fact, the new assumption is implied by the original one. See Sect. 7.4.5.) Similarly to the previous case, we show that under this assumption $D_\mu(M)$ is pseudorandom.

Now, whenever $\mu = 2^{-t}$ for some integer t, we can sample each bit of the noise vector by taking the product of t random bits. Hence, in this case $D_\mu(M)$ is samplable in **NC**^{0} (as we think of μ as a constant). The problem is that our **NC**^{0} procedure which samples $D_\mu(M)$ consumes more bits than it produces (i.e., it consumes $n + t \cdot m$ bits and produces m bits). Hence, we lose the stretch. To solve this, we note that most of the entropy of the seed was not used. Thus, we can gain more output bits by applying a randomness extractor to the seed. To be useful, this randomness extractor should be computable in **NC**^{0}. We construct such an extractor by relying on the construction of an ε-biased generator in **NC**^{0} of [112].

For ease of presentation, we describe our construction in a non-uniform way. We will later discuss a uniform variant of the construction.

7.4.2 The Assumption

Let $m = m(n)$ be an output length parameter where $m(n) > n$, let $\ell = \ell(n)$ be a locality parameter (typically a constant), and let $0 < \mu < 1/2$ be a noise parameter. Let $\mathcal{M}_{m,n,\ell}$ be the set of all $m \times n$ matrices over $\mathbb{F}_2$ in which each row contains

---

[3]Indeed, some impossibility results regarding randomness-efficient **NC**^{0} sampling of the error distribution have recently appeared in [140].

at most ℓ ones. For a matrix $M \in \mathcal{M}_{m,n,\ell}$ denote by $D_\mu(M)$ the distribution of the random m-bit vector

$$Mx + e,$$

where $x \leftarrow U_n$ and $e \in \{0, 1\}^m$ is a random error vector in which each entry is chosen to be 1 with probability μ (independently of other entries), and arithmetic is over $\mathbb{F}_2$. The following assumption is a close variant of a conjecture suggested by Alekhnovich in [5, Conjecture 1].[4]

Intractability Assumption 7.1 *For any $m(n) = O(n)$, and any constant $0 < \mu < 1/2$, there exists a positive integer ℓ, and an infinite family of matrices $\{M_n\}_{n\in\mathbb{N}}$, $M_n \in \mathcal{M}_{m(n),n,\ell}$, such that*

$$D_\mu(M_n) \overset{c}{\equiv} D_{\mu + m(n)^{-1}}(M_n).$$

(Note that since we consider non-uniform distinguishers, we can assume that M_n is public and is available to the distinguisher.)

Remark 7.3 Note that in Assumption 7.1 we do not require $\{M_n\}$ to be polynomial-time computable. We will later present a uniform construction based on the following version of Assumption 7.1. For any $m(n) = O(n)$, any constant $0 < \mu < 1/2$, and any infinite family of $m(n) \times n$ binary matrices $\{M_n\}_{n\in\mathbb{N}}$, if $\{M_n\}$ is expanding then $D_\mu(M_n) \overset{c}{\equiv} D_{\mu + m(n)^{-1}}(M_n)$. This assumption seems likely as argued by Alekhnovich [5, Remark 1].

The following lemma shows that if the distribution $D_\mu(M_n)$ satisfies the above assumption then it is pseudorandom. (The proof is very similar to the proof of [5, Theorem 3.1], and it is given here for completeness.)

Lemma 7.2 *For any polynomial $m(n)$ and constant $0 < \mu < 1/2$, and any infinite family, $\{M_n\}_{n\in\mathbb{N}}$, of $m(n) \times n$ matrices over $\mathbb{F}_2$, if $D_\mu(M_n) \overset{c}{\equiv} D_{\mu + m(n)^{-1}}(M_n)$, then $D_\mu(M_n) \overset{c}{\equiv} U_{m(n)}$.*

Proof Let $m = m(n)$. Let r_n denote the distribution of an m-bit vector in which each entry is chosen to be 1 with probability c/m (independently of other entries) where

---

[4]Our assumption is essentially the same as Alekhnovich's. The main difference between the two assumptions is that the noise vector e in [5] is a random vector of weight exactly $\lceil \mu m \rceil$, as opposed to our noise vector whose entries are chosen to be 1 independently with probability μ. In Sect. 7.4.5 we show that our assumption is implied by Alekhnovich's assumption. Intuitively, the implication follows from the fact that our noise vectors can be viewed as a convex combination of noise vectors of fixed weight. We do not know whether the converse implication holds. Indeed, a distribution D which can be described as a convex combination of distributions $D_1, \ldots, D_n$ may be pseudorandom even if each of the distributions D_i is not pseudorandom.

c is the constant $1/(1 - 2\mu)$. As shown next, we can write

$$D_{\mu+m^{-1}}(M_n) \equiv D_\mu(M_n) + r_n. \tag{7.1}$$

To see this, let $e, e' \in \{0, 1\}^m$ be noise vectors of rate $\mu, \mu + 1/m$ respectively. Then, to prove Eq. (7.1) it suffices to show that $e' \equiv e + r_n$. Indeed, the entries of $e + r_n$ are iid Bernoulli random variables whose success probability is

$$\mu \cdot \left(1 - \left(m(1 - 2\mu)\right)^{-1}\right) + (1 - \mu) \cdot \left(m(1 - 2\mu)\right)^{-1} = \mu + m(n)^{-1}.$$

Now, by Eq. (7.1) and the lemma's hypothesis, we have

$$D_\mu(M_n) \overset{c}{\equiv} D_\mu(M_n) + r_n. \tag{7.2}$$

Let r_n^i be the distribution resulting from summing (over $\mathbb{F}_2^m$) i independent samples from r_n. Let $p(\cdot)$ be a polynomial. Then, by Fact 2.11, we get that

$$D_\mu(M_n) \overset{c}{\equiv} D_\mu(M_n) + r_n^{p(n)}. \tag{7.3}$$

Recall that r_n is a vector of iid Bernoulli random variables whose success probability is $\Theta(1/m)$. Hence, for some polynomial $p(\cdot)$ (e.g., $p(n) = nm$) it holds that

$$r_n^{p(n)} \overset{s}{\equiv} U_{m(n)}. \tag{7.4}$$

(To see this, note that $r_n^{p(n)}$ is a vector of iid Bernoulli random variables whose success probability is, by Fact 7.2, $1/2 \pm (1/2 - \Theta(1/m))^{p(n)} = 1/2 \pm \text{neg}(n)$.) By combining Eqs. (7.3) and (7.4), we have

$$D_\mu(M_n) \overset{c}{\equiv} D_\mu(M_n) + r_n^{p(n)} \overset{s}{\equiv} D_\mu(M_n) + U_{m(n)} \equiv U_{m(n)},$$

and the lemma follows. $\square$

By combining Assumption 7.1 and Lemma 7.2, we get the following.

Proposition 7.1 *Suppose that Assumption 7.1 holds. Then, for any $m(n) = O(n)$, and any constant $0 < \mu < 1/2$, there exists a constant $\ell \in \mathbb{N}$, and an infinite family of matrices $\{M_n\}_{n \in \mathbb{N}}$ where $M_n \in \mathcal{M}_{m(n), n, \ell}$ such that $D_\mu(M_n) \overset{c}{\equiv} U_{m(n)}$.*

Remark 7.4 If the restriction on the density of the matrices M_n is dropped, the above proposition can be based on the conjectured (average case) hardness of decoding a random linear code (cf. [32, 76]). In fact, under the latter assumption we have that $D_\mu(M_n) \overset{c}{\equiv} U_{m(n)}$ for *most* choices of M_n's.

7.4.3 The Construction

From here on, we let $\mu = 2^{-t}$ for some $t \in \mathbb{N}$. Then, we can sample each bit of the error vector e by taking the product of t independent random bits. This naturally gives rise to an **NC**^{0} function whose output distribution is pseudorandom, namely,

$$f_n(x, y) = M_n x + E(y)$$

where

$$x \in \{0, 1\}^n, \qquad y \in \{0, 1\}^{t \cdot m(n)}, \qquad E(y) = \left(\prod_{j=1}^{t} y_{t \cdot (i-1)+j} \right)_{i=1}^{m(n)}. \qquad (7.5)$$

Since $f_n(U_n, U_{t \cdot m(n)}) \equiv D_\mu(M_n)$, the distribution $f_n(U_n, U_{t \cdot m(n)})$ is pseudorandom under Assumption 7.1 (when the parameters are chosen appropriately). Moreover, the locality of f_n is $\ell + t = O(1)$. However, f_n is not a pseudorandom generator as it uses $n + t \cdot m(n)$ input bits while it outputs only $m(n)$ bits. To overcome this obstacle, we note that most of the entropy of y was not "used". Indeed, we use the $t \cdot m(n)$ random bits of y to sample the distribution $E(y)$ whose entropy is only $m(n) \cdot H_2(2^{-t}) < (t+2) \cdot 2^{-t} \cdot m(n)$. Hence, we can apply an *extractor* to regain the lost entropy. Of course, in order to get a PRG in **NC**^{0} the extractor should also be computed in **NC**^{0}. Moreover, to get a linear stretch we should extract almost all of the $t \cdot m(n)$ random bits from y while investing less than m additional random bits. In the following, we show that such extractors can be implemented by using ε-*biased generators*.

First, we show that the distribution of y given $E(y)$ contains (with high probability) a lot of entropy. In the following we let $m = m(n)$.

Lemma 7.3 *Let* $y \leftarrow U_{t \cdot m}$ *and* $E(y)$ *be defined as in Eq.* (7.5). *Denote by* $[y|E(y)]$ *the distribution of* y *given the outcome of* $E(y)$. *Then, except with probability* $e^{-(2^{-t}m)/3}$ *over the choice of* y, *it holds that*

$$H_\infty([y|E(y)]) \ge (1 - \delta(t)) \cdot tm, \qquad (7.6)$$

where $\delta(t) = 2^{-\Omega(t)}$.

Proof We view $E(y)$ as a sequence of m independent Bernoulli trials, each with a probability 2^{-t} of success. Recall that y is composed of m blocks of length t, and that the i-th bit of $E(y)$ equals the product of the bits in the i-th block of y. Hence, whenever $E(y)_i = 1$ all the bits of the i-th block of y equal to 1, and when $E(y)_i = 0$ the i-th block of y is uniformly distributed over $\{0, 1\}^t \setminus \{1^t\}$. Consider the case in which at most $2 \cdot 2^{-t}m$ components of $E(y)$ are ones. By a Chernoff bound, the probability of this event is at least $1 - e^{-(2^{-t}m)/3}$. In this case, y is uniformly distributed over a set of size at least $(2^t - 1)^{(1-2^{-t+1})m}$. Hence, conditioning on the event that at most $2 \cdot 2^{-t}m$ components of $E(y)$ are ones, the min-entropy of $[y|E(y)]$ is at least $m(1 - 2^{-t+1})\log(2^t - 1) \geq tm(1 - \delta(t))$, for $\delta(t) = 2^{-\Omega(t)}$. $\square$

ε-biased generators can be used to extract random bits from distributions that contain sufficient randomness. Extractors based on ε-biased generators were previously used in [56].

Lemma 7.4 ([56, Lemma 4]) *Let $g : \{0, 1\}^n \to \{0, 1\}^s$ be an ε-biased generator, and let X_s be a random variable taking values in $\{0, 1\}^s$ whose min-entropy is at least $(1 - \delta) \cdot s$, for some $\delta \geq 0$. Then,*

$$\left\| (g(U_n) + X_s) - U_s \right\| \leq \varepsilon \cdot 2^{\delta \cdot s/2 - 1/2} ,$$

where vector addition is taken over $\mathbb{F}_2$.

The above lemma follows directly by analyzing the effect of a random step over a Cayley graph whose generator set is an ε-biased set (cf. [79, Lemma 2.3] and [8, 113]).

Mossel et al. [112] constructed an ε-biased generator in nonuniform-$\mathbf{NC}^0_5$ with an arbitrary linear stretch and exponentially small bias.

Lemma 7.5 ([112, Theorem 14]) *For every constant c, there exists an ε-biased generator $g : \{0, 1\}^n \to \{0, 1\}^{cn}$ in nonuniform-$\mathbf{NC}^0_5$ whose bias is at most $2^{-bn/c^4}$ (where $b > 0$ is some universal constant that does not depend on c).*

In Sect. 7.4.4 we provide a uniform version of the above lemma in which the bias is only $2^{-n/\text{polylog}(c)}$. The price we pay is in the locality which grows polylogarithmically with the stretch constant c. (See Theorem 7.4.)

We can now describe our LPRG.

Construction 7.1 *Let t and ℓ be positive integers, and $c, k > 1$ be real numbers that will affect the stretch factor. Let $m = kn$ and let $\{M_n \in \mathcal{M}_{n,m,\ell}\}$ be an infinite family of matrices. Let $g : \{0, 1\}^{tm/c} \to \{0, 1\}^{tm}$ be the ε-biased generator promised by Lemma 7.5. We define the function*

$$G_n(x, y, r) = \big(M_n x + E(y), g(r) + y\big),$$

where $x \in \{0, 1\}^n, y \in \{0, 1\}^{t \cdot m}, r \in \{0, 1\}^{t \cdot m/c}, E(y) = (\prod_{j=1}^{t} y_{t \cdot (i-1)+j})_{i=1}^{m} \cdot$ Thus, $G_n : \{0, 1\}^{n+tm+\frac{tm}{c}} \to \{0, 1\}^{m+tm}$.

Observe that G_n is in nonuniform-$\mathbf{NC}^0$. We show that if the parameters are chosen properly then G_n is an LPRG.

Lemma 7.6 *Under Assumption 7.1, there exist constants $t, \ell \in \mathbb{N}$, constants $c, k > 1$, and a family of matrices $\{M_n \in \mathcal{M}_{n,m,\ell}\}$ such that the function G_n defined in Construction 7.1 is an LPRG.*

Proof Let $k > 1$ be some arbitrary constant and $m = m(n) = kn$. Let c and t be constants such that:

$$c = 2t/(1 - 1/k)$$

and

$$\Delta \stackrel{\text{def}}{=} t\left(\frac{b}{c^5} - \delta(t)\right) > 0, \tag{7.7}$$

where $\delta(\cdot)$ is the negligible function from Eq. (7.6) and b is the bias constant of Lemma 7.5. Such constants c and t do exist since $\delta(t) = 2^{-\Omega(t)}$ while $b/c^5 = \Theta(1/t^5)$. Let $\ell \in \mathbb{N}$ be a constant and $\{M_n \in \mathcal{M}_{n,m,\ell}\}$ be an infinite family of matrices satisfying Assumption 7.1.

First, we show that G_n has linear stretch. The input length of G_n is $n + tm + tm/c = (tk + k/2 + 1/2) \cdot n$. The output length is $(t + 1) \cdot m = (tk + k) \cdot n$. Hence, since $k > 1$, the function G_n has a linear stretch.

Let x, y and r be uniformly distributed over $\{0, 1\}^n$, $\{0, 1\}^{t \cdot m}$ and $\{0, 1\}^{t \cdot m/c}$ respectively. We prove that the distribution $G_{M_n}(x, y, r)$ is pseudorandom. By Fact 2.4 and Lemmas 7.3, 7.4 and 7.5 it holds that

$$\left\| (E(y), y + g(r)) - (E(y), U_{t \cdot m}) \right\| \le e^{-(2^{-t}m)/3} + 2^{-b \cdot (tm/c)/c^4} \cdot 2^{tm \cdot \delta(t)/2 - 1/2}$$

$$\le e^{-(2^{-t}m)/3} + 2^{(-b/c^5 + \delta(t)) \cdot tm}$$

$$\le e^{-(2^{-t}m)/3} + 2^{-\Delta m} = \text{neg}(m) = \text{neg}(n),$$

where the last inequality is due to Eq. (7.7). Therefore, by Fact 2.3 and Proposition 7.1, we get that

$$\left(M_n x + E(y), g(r) + y\right) \stackrel{s}{\equiv} \left(M_n x + E(y), U_{t \cdot m}\right) \equiv \left(D_{2^{-t}}(M_n), U_{t \cdot m}\right) \stackrel{c}{\equiv} (U_m, U_{t \cdot m}),$$

and the lemma follows. $\square$

By the above lemma we get a construction of LPRG in nonuniform-$\mathbf{NC}^0$ from Assumption 7.1. Moreover, by combining the above with Theorem 4.7 we have the following.

Theorem 7.3 *Under Assumption 7.1, there exists an LPRG in nonuniform-$\mathbf{NC}_4^0$.*

Mossel et al. [112] showed that a PRG in nonuniform-$\mathbf{NC}_4^0$ cannot achieve a superlinear stretch. Hence, Theorem 7.3 is essentially optimal with respect to stretch.

Remarks on Theorem 7.3

1. **Uniformity**. Our construction uses two non-uniform advices: (1) a family of good ε-biased generators in $\mathbf{NC}^0$ as in Lemma 7.5; and (2) a family of matrices $\{M_n\}$ satisfying Assumption 7.1. In Sect. 7.4.4 we eliminate the use of the first advice by proving a uniform version of Lemma 7.5. We can also eliminate the

second advice and construct an LPRG in *uniform* $\mathbf{NC}_4^0$ by using an explicit variant of Assumption 7.1. In particular, we follow Alekhnovich (cf. [5, Remark 1]) and conjecture that any family of matrices $\{M_n\}$ that represent graphs with good expansion satisfies Assumption 7.1. Hence, our construction can be implemented by using an explicit family of asymmetric constant-degree bipartite expanders such as the one given in [44, Theorem 7.1].

2. **PRG with constant input locality**. A variant of our construction also gives a PRG G in $\mathbf{NC}^0$ in which each input bit affects a constant number of output bits. Namely, G enjoys constant output locality and constant *input* locality at the same time. This can be done by instantiating Construction 7.1 with a family of matrices $\{M_n\}$ in which each row and each *column* contain a constant number of 1's, and, in addition, employing an ε-biased generator with constant input locality and constant output locality. It turns out that the instantiation proposed in the previous item (which is based on [44, Theorem 7.1] and Theorem 7.4) satisfies these conditions. Thus, assuming that Assumption 7.1 holds whenever $\{M_n\}$ is expanding, we get a (uniform) LPRG with constant output locality and constant input locality. In Chap. 8, we will construct PRGs with better (optimal) input and output locality under a seemingly weaker assumption (namely, the intractability of decoding random linear code). However, the construction of Chap. 8 is limited to *sublinear* stretch.

3. **The stretch of the construction**. Our techniques do not yield a PRG with *superlinear* stretch in $\mathbf{NC}^0$. To see this, consider a variant of Assumption 7.1 in which we allow $m(n)$ to be superlinear. If we let $\mu(n)$ be a constant, then, by information-theoretic arguments, we need $\Omega(m(n))$ random bits to sample the noise vector (i.e., the entropy of the noise vector is $\Omega(m(n))$), and so we get only linear stretch. On the other hand, if we set $\mu(n)$ to be subconstant, then the noise distribution cannot be sampled in $\mathbf{NC}^0$ (as any bit of an $\mathbf{NC}^0$-samplable distribution depends on a constant number of random bits). This problem can be bypassed by extending Assumption 7.1 to alternative noise models in which the noise is not independently and identically distributed. However, it is not clear how such a modification affects the hardness assumption. (Also note that we do not know how to reduce the locality of a superlinear PRG in $\mathbf{NC}^0$ while preserving its superlinear stretch. In particular, applying the transformations of Chap. 4 to such a PRG will result in a *linear* PRG with locality 4.)

7.4.4 ε-Biased Generators in Uniform $\mathbf{NC}^0$

In [112, Theorem 14], Mossel et al. described a non-uniform construction of an ε-biased generator with locality 5, an arbitrary linear stretch cn and bias $\varepsilon = 2^{-\Omega(n/c^4)}$.[5] We generalize their construction and provide a complementary result

---

[5]In fact, cn can be slightly superlinear.

which gives a better tradeoff between the bias and stretch and allows a uniform implementation. However, the locality of our construction grows with the stretch constant.

Theorem 7.4 *For every sufficiently large constant c, there exists an ε-biased generator $g : \{0, 1\}^n \to \{0, 1\}^{cn}$ in uniform **NC**^{0} whose bias is $\varepsilon = 2^{-n/\text{polylog}(c)}$ and its output locality is $\ell = \text{polylog}(c)$. Moreover, the input locality of g is constant (which depends on c).*

As in [112], our generator is obtained by XORing the outputs of two functions: a generator $g^{(s)}$ which is robust against linear functions that involve a small number of output bits ("small tests") and a generator $g^{(l)}$ which is robust against linear functions that involve a large number of output bits ("large tests"). More precisely, for a random variable $X = (X_1, \ldots, X_n)$ ranging over $\{0, 1\}^n$, a set $S \subseteq \{1, \ldots, n\}$, and an integer $0 < k \le n$, we define

$$\text{bias}_S(X) \overset{\text{def}}{=} \left| \Pr\left[\bigoplus_{i \in S} X_i = 0 \right] - \frac{1}{2} \right|,$$

$$\text{bias}_k(X) \overset{\text{def}}{=} \max_{S \subseteq \{1,\ldots,n\}, |S|=k} \text{bias}_S(X),$$

$$\text{bias}(X) \overset{\text{def}}{=} \max_{0<k\le n} \text{bias}_k(X) = \max_{S \subseteq \{1,\ldots,n\}, S \ne \emptyset} \text{bias}_S(X).$$

Then, we prove Theorem 7.4 by using the following two lemmas (whose proofs are postponed to Sects. 7.4.4.1 and 7.4.4.2):

Lemma 7.7 (Generator against small tests) *For every constant c, there exists a function $g^{(s)} : \{0, 1\}^n \to \{0, 1\}^{cn}$ in uniform **NC**$^0_{\text{polylog}(c)}$ such that for sufficiently large n's and every $0 < k \le n/\text{polylog}(c)$, we have $\text{bias}_k(g^{(s)}(U_n)) = 0$. Moreover, the input locality of $g^{(s)}$ is constant (which depends on c).*

Lemma 7.8 (Generator against large tests) *For every constant c, there exists a function $g^{(l)} : \{0, 1\}^n \to \{0, 1\}^{cn}$ in uniform **NC**$^0_{O(\log(c))}$ such that for sufficiently large n's and every $k \in \{1, \ldots, cn\}$, we have $\text{bias}_k(g^{(l)}(U_n)) \le 2^{-k/5}$. Moreover, the input locality of $g^{(s)}$ is constant (which depends on c).*

Given these two lemmas we can prove Theorem 7.4.

Proof of Theorem 7.4 Let c be a constant. Let $g^{(s)} : \{0, 1\}^n \to \{0, 1\}^{2cn}$ and $g^{(l)} : \{0, 1\}^n \to \{0, 1\}^{2cn}$ be the generators promised by Lemmas 7.7 and 7.8 (instantiate with the constant 2c). Then, the function $g(x, y) = g^{(s)}(x) \oplus g^{(l)}(y)$ satisfies Theorem 7.4. To see this, observe that for any *independent* random variables X and Y and any non-uniform statistical test T, the success probability of T on the random variable $X \oplus Y$ is not larger than its success probability on X (or Y). $\square$

7.4.4.1 Proof of Lemma 7.7

Let M be an $m \times n$ matrix over $\mathbb{F}_2$ such that every subset of k rows of M are linearly independent. Then, it is well known that the function $f : \{0, 1\}^n \rightarrow \{0, 1\}^m$ that maps x into $M \cdot x$ is a k-wise independent generator (cf. [7]). That is, for every $0 < j \leq k$, we have $\mathrm{bias}_j(f(U_n)) = 0$. If each row of M contains at most ℓ ones then the function f is in nonuniform-$\mathbf{NC}_\ell^0$. It turns out that there exists a (uniform) family of such matrices whose parameters match the parameters of Lemma 7.7. Specifically, we use the following result which is a corollary of [44, Theorem 7.1].

Lemma 7.9 ([44]) *For every sufficiently large constant c there exists a family of matrices $\{M_n\}_{n \in \mathbb{N}}$ such that*

- M_n *is a $cn \times n$ matrix over $\mathbb{F}_2$.*
- *Every row of M_n has at most* polylog(c) *ones.*
- *Every column of M_n has at most $d(c)$ ones for some function $d(\cdot)$.*
- *Every subset of $k = n/$polylog(c) rows of M_n is linearly independent.*
- M_n *can be constructed in time* poly(n).

Hence, the generator for small tests can be defined as $g^{(s)}(x) = M_n \cdot x$.

7.4.4.2 Proof of Lemma 7.8

We will need the following standard claim that can be proved via the probabilistic method (see [71, Lecture 8, Proposition 2.1]).

Claim 7.1 *For sufficiently large n, there exists an ε-biased generator $f : \{0, 1\}^n \rightarrow \{0, 1\}^{2n/2}$ whose bias is $\varepsilon = 2^{-n/4}$.*

We can now prove Lemma 7.8. Let c be the desired stretch constant. Let $\ell = 4 \log c$. Let $m = 2^{\ell/2}$ and $f : \{0, 1\}^\ell \rightarrow \{0, 1\}^m$ be an ε-biased generator whose bias is $\varepsilon = 2^{-\ell/4}$ as promised by Claim 7.1. (Since c is a constant, such f can be found by using exhaustive search.) Our generator will partition its n-bit input x into $b = \lfloor n/\ell \rfloor$ blocks $x^{(1)}, \ldots, x^{(b)}$ of length ℓ each. Then, the generator will apply f to each block separately, and concatenate the result. Namely, $g^{(l)}(x) \overset{\mathrm{def}}{=} (f(x^{(1)}), \ldots, f(x^{(b)}))$. The output locality of $g^{(l)}$ is ℓ, its input locality is $m = 2^{\ell/2} = c^2$ and its output length is $bm = \lfloor \frac{c^2 n}{4 \log c} \rfloor$ which is larger than cn for sufficiently large c.

We now analyze the bias of $g^{(l)}$. To simplify notation, we index the outputs of $g^{(l)}$ by pairs (j, i) and let $g_{j,i}^{(l)}(x) = f_i(x^{(j)})$ (where $1 \leq j \leq b$, $1 \leq i \leq m$ and $f_i(x)$ denotes the i-th output bit of $f(x)$). Let $S \subseteq \{1, \ldots, b\} \times \{1, \ldots, m\}$ be a linear test of cardinality k. Let S_j be the restriction of S to the indices of the j-th block, i.e., $S_j = \{i : (j, i) \in S\}$. Then, $S_1, \ldots, S_b$ is a partition of S. Let $T = \{i : S_i \neq \emptyset\} \subseteq$

$\{1, \ldots, b\}$. Hence, for $x \leftarrow U_n$, we have

$$\text{bias}_S\big(g^{(l)}(x)\big) = \text{bias}\Big(\bigoplus_{j \in T} \bigoplus_{i \in S_j} f_i(x^{(j)})\Big).$$

Since f is an ε-biased generator, for each $j \in T$ we have that $\text{bias}(\bigoplus_{i \in S_j} f_i(x^{(j)}))$
$\leq \varepsilon$. Since $g^{(l)}(x)$ is partitioned into blocks of length ℓ, the test S contains output
bits coming from at least k/ℓ different blocks and so $|T| \geq k/\ell$. Thus we can use
Fact 7.2 to upper bound $\text{bias}_S(g^{(l)}(x))$ by

$$\frac{1}{2}(2\varepsilon)^{k/\ell} \leq \frac{1}{2}\big(2^{-\ell/4+1}\big)^{k/\ell} \leq \frac{1}{2}\big(2^{-\ell/5}\big)^{k/\ell} \leq 2^{-k/5},$$

as required. $\square$

7.4.5 Alekhnovich's Assumption Implies Assumption 7.1

We show that Alekhnovich's Assumption [5, Conjecture 2, Remark 1] implies Assumption 7.1. The main difference between the two assumptions is that the noise vector e in [5] is a random vector of weight exactly $\lceil \mu m \rceil$, as opposed to our noise vector whose entries are chosen to be 1 independently with probability μ. The implication follows from the fact that our noise vectors can be viewed as a convex combination of noise vectors of fixed weight. We give the details below.

Recall that for an $m \times n$ matrix M we let $d_\mu(M)$ denote the distribution of $M \cdot x + e$, where x is a random n-bit vector and e is a noise vector which is uniformly distributed over all m-bits vectors whose Hamming weight is $\mu \cdot m$. The distribution $D_\mu(M) \stackrel{\text{def}}{=} M \cdot x + e$ is analogous to $d_\mu(M)$, except that each entry of the noise vector e is chosen to be 1 with probability μ (independently of other entries).

Intractability Assumption 7.2 (Alekhnovich's assumption) *For any $m(n) = O(n)$, there exists an infinite family of matrices $\{M_n\}_{n \in \mathbb{N}}$, $M_n \in \mathcal{M}_{m(n),n,3}$, such that for any constant $0 < \mu_0 < 1/2$, and function $\mu(n)$ that satisfies $\mu_0 < \mu(n) < 1/2$ for every n, it holds that*

$$d_{\mu(n)}(M_n) \stackrel{c}{\equiv} d_{\mu(n)+m(n)^{-1}}(M_n).$$

Fix a matrix family $\{M_n\}_{n \in \mathbb{N}}$ of size $m(n) \times n$ where $m(n)$ is an integer valued function. We will prove that Assumption 7.2 instantiated with the family $\{M_n\}_{n \in \mathbb{N}}$ implies Assumption 7.1 instantiated with the same family of matrices. To do this we use the following two intermediate assumptions.

Intractability Assumption 7.3 *For any constant $0 < \mu_0 < 1/2$, and function $\mu(n)$ that satisfies $\mu_0 < \mu(n) < 1/2$ for all n's, $d_{\mu(n)}(M_n) \stackrel{c}{\equiv} U_{m(n)}$.*

Intractability Assumption 7.4 *For any constant* $0 < \mu < 1/2$, *we have* $D_\mu(M_n) \overset{c}{\equiv} U_{m(n)}$.

In [5, Theorem 3.1] it is shown that Assumption 7.2 implies Assumption 7.3. Hence to prove that Assumption 7.2 implies Assumption 7.1 it suffices to show that: (1) Assumption 7.3 implies Assumption 7.4; and (2) Assumption 7.4 implies Assumption 7.1.

Lemma 7.10 *Assumption* 7.3 *implies Assumption* 7.4.

Proof Suppose that Assumption 7.4 does not hold. Then, for some constant $0 < \mu < 1/2$, the distribution $D_\mu(M_n)$ is not pseudorandom. That is, there exists a polynomial-size circuit family $\{A_n\}$ and a polynomial $q(\cdot)$ such that

$$\Pr[A_n(D_\mu(M_n)) = 1] - \Pr[A_n(U_{m(n)}) = 1] > 1/q(n), \qquad (7.8)$$

for infinitely many n's. We will show that, for some constant $0 < \hat{\mu}_0 < 1/2$, and function $\hat{\mu}(n)$ that satisfies $\hat{\mu}_0 < \hat{\mu}(n) < 1/2$, Assumption 7.3 is violated. Namely, $\Pr[A_n(d_{\hat{\mu}(n)}(M_n)) = 1] - \Pr[A_n(U_{m(n)}) = 1] > 1/q'(n)$ for some polynomial $q'(\cdot)$ and infinitely many n's.

Fix some n for which Eq. (7.8) holds, and let $m = m(n)$. Let $p \overset{\text{def}}{=} \Pr[A_n(D_\mu(M_n)) = 1]$ and $p(k) \overset{\text{def}}{=} \Pr[A_n(d_{k/m}(M_n)) = 1]$ for $0 \le k \le m$. Let $e \in \{0,1\}^m$ be a random error vector in which each entry is chosen to be 1 with probability μ (independently of other entries) and let $t(k)$ be the probability that e contains exactly k ones. We can think of the distribution of e as the outcome of the following process: first choose $0 \le k \le m$ with probability $t(k)$, then choose a random noise vector of weight k. Hence, we can write,

$$p = \sum_{k=0}^{m} p(k) \cdot t(k).$$

Let $\varepsilon > 0$ be a constant for which $\mu \cdot \varepsilon < 1/2$. Then, by a Chernoff bound, it holds that

$$\sum_{k < (1-\varepsilon)\cdot\mu m} t(k) + \sum_{k > (1+\varepsilon)\cdot\mu m} t(k) = \Pr\left[\left|\sum_{i=1}^{m} e_i - \mu m\right| > \varepsilon \cdot \mu m\right] < 2e^{-\varepsilon^2 \mu m/3}.$$

Hence,

$$\sum_{(1-\varepsilon)\cdot\mu m \le k \le (1+\varepsilon)\cdot\mu m} p(k) \cdot t(k) > p - 2e^{-\varepsilon^2 \mu m/3}.$$

Thus, by an averaging argument, there exists some $(1-\varepsilon) \cdot \mu m \le k \le (1-\varepsilon) \cdot \mu m$ for which

$$p(k) > p - 2e^{-\varepsilon^2 \mu m/3}.$$

Let $\hat{\mu}(n)$ be k/m and let $\hat{\mu}_0$ be the constant $(1 - \varepsilon) \cdot \mu m/2$. Then, by Eq. (7.8), we have

$$\Pr\big[A_n\big(d_{\hat{\mu}(n)}(M_n)\big) = 1\big] - \Pr\big[A_n(U_{m(n)}) = 1\big] > 1/q(n) - 2e^{-\varepsilon^2 \mu m/3} > 1/q'(n),$$

where $q'(\cdot)$ is a polynomial. This completes the proof since $\hat{\mu}_0 < \hat{\mu}(n) < 1/2$ for every n.   □

It is left to prove the following lemma.

Lemma 7.11 *If Assumption 7.4 holds then Assumption 7.1 also holds with respect to* $\{M_n\}_{n \in \mathbb{N}}$.

Proof As shown in the proof of Lemma 7.2 we can write $D_{\mu+m-1}(M_n) \equiv D_\mu(M_n) + r_n$, where r_n denotes the distribution of an m-bit vector in which each entry is chosen to be 1 with probability c/m (independently of other entries) for some constant c. Hence, by two invocations of Assumption 7.4, we have

$$D_{\mu+m-1}(M_n) \equiv D_\mu(M_n) + r_n \overset{c}{\equiv} U_{m(n)} + r_n \equiv U_{m(n)} \overset{c}{\equiv} D_\mu(M_n). \qquad \square$$

7.5 The Necessity of Expansion

As pointed out in Sect. 7.4, our construction of LPRG makes use of expander graphs. This is also the case in several constructions of "hard functions" with low locality (e.g., [5, 69, 112]). We argue that this is not coincidental, at least in the case of PRGs. Namely, we show that the input-output graph of any LPRG in $\mathbf{NC}^0$ enjoys some expansion property. (In fact, this holds even in the case of ε-biased generators.) Then, we use known lower bounds for expander graphs to rule out the possibility of exponentially strong PRG with superlinear stretch in $\mathbf{NC}^0$. These results are discussed from a wider perspective in Sect. 7.5.2. We start with the technical results.

7.5.1 Results

For a function $G : \{0, 1\}^n \to \{0, 1\}^s$, we define the input-output graph $H_G = ((\text{Out} = [s], \text{In} = [n]), E)$ to be the bipartite graph whose edges correspond to the input-output dependencies in G; that is, (i, j) is an edge if and only if the i-th output bit of G depends on the j-th input bit. When G is a function family, H_G denotes a graph family.

Proposition 7.2 *Let* $G : \{0, 1\}^n \to \{0, 1\}^{s(n)}$ *be a PRG. Then, the graph (family)* $H_G = ((\text{Out} = [s(n)], \text{In} = [n]), E)$ *enjoys the following output expansion property: for every constant c and sufficiently large n, every set of output vertices $T \subseteq \text{Out}$ whose size is at most $c \log n$ touches at least $|T|$ input vertices.*

Proof Assume towards a contradiction that there exists a small set T of output vertices that touches fewer than $|T|$ input vertices. Let $G_T(\cdot)$ be the restriction of G to the output bits of T. Then, the function $G_T(\cdot)$ cannot be onto as it depends on less than $|T|$ input bits. Therefore, there exists a string $z \in \{0, 1\}^{|T|}$ such that $\Pr[G_T(U_n) = z] = 0$. Hence, a (non-uniform) distinguisher which given $y \in \{0, 1\}^{s(n)}$ checks whether $y_T = z$, distinguishes between $G(U_n)$ and $U_{s(n)}$ with advantage $2^{-c \log n} = 1/n^c$, in contradiction to the pseudorandomness of G. $\square$

More generally, if G is ε-hard (i.e., cannot be broken by any efficient adversary with advantage ε), then every set of $t \leq \log(1/\varepsilon)$ output vertices touches at least t input vertices. This claim also extends to the case of ε-biased generators.

Proposition 7.3 *Let $G : \{0, 1\}^n \to \{0, 1\}^s$ be an ε-biased generator. Then, every set of $t \leq \log(1/\varepsilon)$ output vertices in H_G touches at least t input vertices.*

Proof Assume towards a contradiction that there exists a set T of output vertices of size $t \leq \log(1/\varepsilon)$ that touches fewer than t input vertices. Then $G_T(U_n) \not\equiv U_t$. Therefore, there exists a linear function $L : \mathbb{F}_2^t \to \mathbb{F}_2$ that distinguishes between $G_T(U_n)$ and U_t. Namely, $|\Pr[L(G_T(U_n)) = 1] - \Pr[L(U_t) = 1]| \neq 0$. Since the distribution $G_T(U_n)$ is sampled by fewer than t random bits, the distinguishing advantage of L is larger than $2^{-t} \geq \varepsilon$, and so G is not ε-biased in contradiction to the hypothesis. $\square$

The above propositions show that when G is an ε-hard PRG (or even ε-biased generator), the bipartite graph $H_G = ((\text{Out} = [s(n)], \text{In} = [n]), E)$ enjoys some output expansion property. Radhakrishnan and Ta-Shma [124] obtained some lower bounds for such graphs.

Proposition 7.4 ([124, Theorem 1.5]) *Let $H = ((V_1 = [s], V_2 = [n]), E)$ be a bipartite graph in which every set $S \subseteq V_1$ of cardinality k touches at least m vertices from V_2. Then, the average degree of V_1 is at least $\Omega(\frac{\log(s/k)}{\log(m/n)})$.*

By combining this lower bound with the previous propositions we derive the following limitation on the strength of PRGs with superlinear stretch in $\mathbf{NC}^0$.

Corollary 7.3 *Let $G : \{0, 1\}^n \to \{0, 1\}^s$ be a 2^{-t}-hard PRG (or 2^{-t}-biased generator). Then, the locality of G is at least $\Omega(\frac{\log(s/t)}{\log(n/t)})$. In particular, there is no $2^{-\Omega(n)}$-hard PRG, or even a $2^{-\Omega(n)}$-biased generator, with superlinear stretch in $\mathbf{NC}^0$.*

7.5.2 Discussion

To put the above results in context, some background on unbalanced bipartite expanders is needed. Consider a bipartite graph $H = ((\text{Out} = [s], \text{In} = [n]), E)$ in

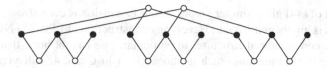

Fig. 7.1 The trivial "slightly" unbalanced expander graph H with $n = 2$ and $m = 3$. Black circles denote output vertices while empty circles denote input vertices

which each of the output vertices is connected to at most d inputs. Recall that H is a (K, α)-expander if every set of output vertices S of size smaller than K has at least $\alpha \cdot |S|$ input neighbors. We say that the expander is unbalanced if $s > n$. Unbalanced expanders have had numerous applications in computer science (see details and references in [44]). Today, there are only two such constructions [44, 133]. Ta-Shma et al. [133] considered the highly unbalanced case in which $n < o(s)$. They constructed a (K, α)-expander with degree $d = \text{polylog}(s)$, expansion threshold $K < s^\varepsilon$ and almost optimal expansion factor $\alpha = (1 - \delta)d$, where $\delta > 0$ is an arbitrary constant. Capalbo et al. [44] present a construction for the setting in which n is an (arbitrary) constant fraction of s (i.e., $s = n + \Theta(n)$). They construct a (K, α)-expander with (nearly) optimal parameters; namely, the degree d of the graph is constant, and its expansion parameters are $K = \Omega(s)$ and $\alpha = (1 - \delta)d$, where $\delta > 0$ is an arbitrary constant.

In Sect. 7.5.1 we showed that if $G : \{0, 1\}^n \to \{0, 1\}^s$ is a PRG then its input-output graph $H_G = ((\text{Out} = [s], \text{In} = [n]), E)$ is an $(\omega(\log n), 1)$-expander. This property is trivial to satisfy when the output degree of H_G is unbounded (as in standard constructions of PRGs in which every output bit depends on all the input bits). It is also easy to construct such a graph with constant output degree when $s(n)$ is not much larger than n (as in the $\mathbf{NC}^0$ constructions of Chap. 4).

To see this, consider the following bipartite graph. First, let $C = ((O, I), D)$ be a bipartite graph over $[2n + 1]$ whose output vertices are the odd integers, its input vertices are the even integers, and its edges correspond to pairs of consecutive integers, i.e., $O = \{1, 3, \ldots, 2n + 1\}$, $I = \{2, 4, \ldots, 2n\}$, and D contains the edges $(1, 2), (2, 3), \ldots, (2n, 2n + 1)$. That is, C is a chain of length $2n + 1$. Let $m > n$. Take m disjoint copies of C, and let O_i (resp. I_i) be the set of output (resp. input) vertices of the i-th copy. In addition, add n input vertices $I_0 = [n]$ and match them to the first n vertices of each of the output clusters (i.e., connect the j-th vertex of I_0 to the vertex $2j - 1$ of each O_i). Let $H = ((\text{Out} = O_1 \cup \cdots \cup O_m, \text{In} = I_1 \cup \cdots \cup I_m \cup I_0), E)$. (See Fig. 7.1.) Clearly, H has $m(n + 1)$ output vertices, $mn + n$ input vertices, and each output vertex is connected to at most 3 inputs. It is not hard to verify that H is $(n^2, 1)$-expanding. However, the number of outputs is only slightly larger than the number of inputs; i.e., $|\text{Out}| - |\text{In}| = m - n < m$ which is sublinear in $|\text{In}|$ when n is non-constant.

However, when the locality of the pseudorandom generator G is constant and the stretch is linear, H_G is a sparse bipartite graph having n input vertices, $s(n) = n + \Omega(n)$ output vertices, and a constant output degree. It seems that it is not trivial to explicitly construct such a graph that achieves $(\omega(\log n), 1)$-expansion. (Indeed, the construction of [44] gives similar graphs whose expansion is even stronger, but this construction is highly non-trivial.) Hence, any construction of LPRG in $\mathbf{NC}^0$ defines a non-trivial combinatorial structure. In particular, one cannot hope that "simple" *deterministic* transformations, such as those given in Chap. 4, will yield LPRGs in $\mathbf{NC}^0$.

Note that an exponentially strong PRG (or exponentially strong ε-biased generator) with linear stretch gives an $(\Omega(n), 1)$-expander graph whose output size grows linearly with its input size. Indeed, the exponentially strong ε-biased generator of [112] is based on a similar (but slightly stronger) unbalanced expander. The above argument shows that such an ingredient is necessary.

7.6　Addendum: Notes and Open Questions

7.6.1　Pseudorandomness of Random Local Functions

Alekhnovich's conjecture (Conjecture 1 in [5]) is closely related to Goldreich's ensemble of *random local functions* [69] (see Chap. 6, Assumption 6.1). Recall that for a d-sparse matrix $M \in \{0, 1\}^{m \times n}$ and a d-local predicate $Q : \{0, 1\}^d \rightarrow \{0, 1\}$, we defined the function $f_{M,Q} : \{0, 1\}^n \rightarrow \{0, 1\}^m$ via the mapping

$$x \mapsto \left(Q(x_{M_1}), \ldots, Q(x_{M_m}) \right),$$

where x_{M_i} is the restriction of x to the support of the i-th row of M. For output-length parameter $m = m(n)$, we let $\mathcal{F}_{Q,n,m}$ denote the collection of functions $f_{M,Q}$ where M is a random d-sparse matrix chosen from $\mathcal{M}_{m(n),n,d}$.

Alekhnovich essentially considers the case where M is a random 3-sparse matrix $M \leftarrow \mathcal{M}_{m(n)=O(n),n,3}$ and the predicate Q is *randomized*; i.e., Q computes the μ-noisy XOR predicate $\oplus_\mu$ defined by

$$\oplus_\mu(w_1, w_2, w_3) = \begin{cases} w_1 \oplus w_2 \oplus w_3 & \text{with probability } 1 - \mu, \\ w_1 \oplus w_2 \oplus w_3 \oplus 1 & \text{with probability } \mu, \end{cases}$$

where in each invocation of the predicate the noise is freshly chosen at random.

Conjecture 1 of Alekhnovich [5] can be formulated as saying that this collection is somewhat unpredictable, i.e., given $M \leftarrow \mathcal{M}_{m(n)=O(n),n,3}$ and $f_{M,Q}(U_n)$ it is hard to predict the i-th bit of $f_{M,Q}(U_n)$ based on its prefix.[6] Our construction (Sect. 7.4) essentially shows how to derandomize the internal randomness of the

---

[6]The original formulation asserts that, for a random matrix $M \leftarrow \mathcal{M}_{m(n)=O(n),n,3}$, the pair $(M, y = f_{M,Q}(U_n))$ is indistinguishable from the pair (M, y') where y' is a perturbed version of y in which a single bit in a random location is flipped. This implies that the distribution $f_{M,Q}(U_n)$ is weakly unpredictable.

predicate via an NC^0 extractor which, in turn, is based on an NC^0 small biased generator.

In the last few years, several works have studied the pseudorandomness properties of the collection $\mathcal{F}_{Q,n,m}$. In [11] it was shown that for a "good" choice of the predicate Q, a random function from the collection $\mathcal{F}_{Q,n,m=n^{1+\delta}}$ is likely to be an ε-bias generator. A full classification of good predicates (at the form of a dichotomy theorem) was established in [12]. (These results strengthen the results of [112] which show that *some* members of $\mathcal{F}_{Q,n,m}$ are ε-biased.)

In light of the above results, one may conjecture that the collection $\mathcal{F}_{Q,n,m}$ is cryptographically pseudorandom even for large m. Indeed in [11] it is was shown that, for the special case of the noisy XOR predicate $\oplus_\mu$, one-wayness implies weak pseudorandomness. Specifically, if $\mathcal{F}_{\oplus_\mu,n,m=O(n \log n)}$ is one-way then $\mathcal{F}_{\oplus_p,n,m=O(n)}$ is "somewhat" pseudorandom, as conjectured by Alekhnovich.

The above result was extended in [10] to the case of general predicates. Specifically, it was shown that if $\mathcal{F}_{Q,n,m}$ is one-way then $\mathcal{F}_{Q,n,m'<m}$ has a large amount of "pseudoentropy", or even pseudorandomness if the predicate satisfies some techni cal properties. These results have lead to strong inapproximability results for the densest sub-hypergraph problem, and to new constructions of local PRGs with large stretch. In particular, it was shown how to construct (collections) of linear stretch PRGs in NC^0 based on the one-wayness of $\mathcal{F}_{Q,n,m=O(n)}$, and collections of polynomial-stretch PRGs in NC^0 with inverse polynomial distinguishing advantage based on the one-wayness of $\mathcal{F}_{Q,n,m=n^{1+\delta}}$. The latter result also gives rise to polynomial-stretch PRGs with standard (negligible) distinguishing advantage and super-constant locality, e.g., $\log^*(n)$. The existence of polynomial-stretch PRGs with constant locality and negligible distinguishing advantage remains an interesting open question. A positive answer would lead to secure computation protocols with optimal computational overhead [95].

7.6.2 NC^0 Randomness Extractors and NC^0 Sources

Randomness extractors computable in NC^0 play an important role in this chapter. Formally, a (k, δ) randomness extractor is a function Ext which maps a seed $s \in \{0, 1\}^\sigma$ and a string $x \in \{0, 1\}^n$ into a string $y \in \{0, 1\}^m$ such that for any source X with min-entropy at least k, and a uniformly chosen seed $s \leftarrow \{0, 1\}^\sigma$, the random variable $\text{Ext}_s(X)$ is δ-close (in statistical distance) to the uniform distribution over $\{0, 1\}^m$. In this chapter we described a construction of NC^0 randomness extractor with "linear parameters": min-entropy $k = (1 - \varepsilon)n$, error $\delta = 2^{-\Omega(n)}$, seed length $\sigma = (1 - \varepsilon')n$ and output length $m = n$.

By using the randomized encoding machinery, it is also possible to obtain NC^0 extractors for arbitrary sources with sublinear amount of min-entropy (and negligible error probability) while keeping the entropy loss optimal, i.e., $m = k + \sigma - 2 \log(1/\delta) - O(1)$. However, this is done at the expense of using a *long* seed whose length is (polynomially) larger than the source's length. The construction is obtained by taking a (k, δ) extractor $\text{Ext}(s, x)$ with an optimal entropy loss which is computable in $\mathbf{NC}^1$ (e.g., based on the leftover hashing lemma [86]) and compiling it into a perfect randomized encoding $\hat{\mathrm{Ext}}(s, x; r)$ via the encoding from Chap. 4. It is not hard to show that if the extra randomness r is treated as part of the random seed (and is kept secret) the function $\mathrm{Ext}'((s, r), x) = \hat{\mathrm{Ext}}(s, x; r)$ is also a (k, δ) extractor. (The argument is identical to the one used in the context of pseudo-random generators in Lemma 4.7, except that the distinguisher is computationally unbounded.) Furthermore, since the encoding is stretch preserving, the entropy loss remains the same.

Note that the new part of the seed r must remain private and cannot be output by the extractor even if the original extractor outputs the seed σ (i.e., Ext is a so-called *strong extractor*). In fact, it is not hard to see that any $\mathbf{NC}^0$ extractor must use a secret seed. Indeed, one can easily predict the first output bit of an $\mathbf{NC}^0_d$ extractor $\mathrm{Ext}(s, x)$ given the seed s and the value of the d input bits $(x_{i_1}, \ldots, x_{i_d})$ which influence the first output. Hence, even a single random bit is impossible to (strongly) extract from the source X which is uniformly distributed over $\{x \in \{0, 1\}^n | x_{i_1} = \cdots = x_{i_d} = 0\}$, and therefore has $n - d$ bits of min-entropy.

More refined upper and lower bounds regarding randomness extraction with low locality are obtained in [37]. An orthogonal research direction addresses the task of extracting randomness from *sources* computed by $\mathbf{NC}^0$ circuits [54, 139]. Finally, the randomness complexity of low-depth samplers (for some concrete distributions) is studied in [140].

Chapter 8
Cryptography with Constant Input Locality

Abstract So far we studied the possibility of implementing cryptographic tasks using functions with constant *output* locality. In this chapter we turn to the dual question of constant *input* locality. In particular, we ask: Which cryptographic primitives (if any) can be realized by functions in which every bit of the *input* influences only a constant number of bits of the output? We prove several positive and negative results that together form the following characterization. Essentially, primitives that require some form of non-malleability (e.g., digital signatures) *cannot* be realized with constant input locality, while primitives that require secrecy (e.g., encryption schemes) can be implemented with constant input locality. Some of our constructions enjoy both constant input locality and constant output locality, giving rise to cryptographic hardware that has constant-depth, constant fan-in and constant *fan-out*.

8.1 Introduction

In the previous chapters we showed that, under standard cryptographic assumptions, most cryptographic primitives can be realized using functions with constant *output* locality, namely ones in which every bit of the *output* is influenced by a constant number of bits from the input. In this chapter we study the complementary question of implementing cryptographic primitives by functions in which each *input* bit affects only a constant number of output bits. This was not settled by the previous chapters. This natural question can be motivated from several distinct perspectives:

- **Theoretical examination of a common practice**. A well-known design principle for practical cryptosystems asserts that each input bit must affect many output bits. This principle is sometimes referred to as the Confusion/Diffusion or Avalanche property. It is easy to justify this principle in the context of block-ciphers (which are theoretically modeled as pseudorandom functions or permutations), but is it also necessary in other cryptographic applications (e.g., probabilistic encryption)?
- **Hardware perspective**. Unlike NC^0 functions, functions with both constant input locality and constant output locality can be computed by constant depth circuits with bounded fan-in and *bounded fan-out*. Hence, the parallel-time complexity of such functions is constant in a wider class of implementation scenarios.

B. Applebaum, *Cryptography in Constant Parallel Time*,
Information Security and Cryptography,
DOI 10.1007/978-3-642-17367-7_8, © Springer-Verlag Berlin Heidelberg 2014

- **Complexity theoretic perspective**. One can state the existence of cryptography in NC^0 in terms of average-case hardness of Constraint Satisfaction Problems in which each constraint involves a constant number of variables (k-CSPs). The new question can therefore be formulated in terms of k-CSPs with bounded occurrences of each variable. It is known that NP hardness and inapproximability results can be carried from the CSP setting to this setting [48, 121], hence it is interesting to ask whether the same phenomenon occurs with respect to cryptographic hardness as well.

Motivated by the above, we would like to understand which cryptographic tasks (if any) can be realized with constant input *and* output locality, or even with constant input locality alone. Another question considered in this chapter is that of closing the (small) gap between positive results for cryptography with locality 4 and the impossibility of cryptography with locality 2.

8.1.1 Results

We provide an almost full characterization of the cryptographic tasks that can be realized by functions with constant input locality. On the negative side, we show that primitives which require some form of non-malleability (e.g., signatures, MACs, non-malleable encryption schemes) *cannot* be realized with constant (or, in some cases, even logarithmic) input locality.

On the positive side, assuming the intractability of some problems from the domain of error correcting codes, we obtain constructions of pseudorandom generators, commitments, and semantically secure public-key encryption schemes with constant input locality and constant output locality. In particular, we obtain the following results:

- For PRGs, we answer simultaneously both of the above questions. Namely, we construct a collection[1] of PRGs whose output locality and input locality are both 3. We show that this is optimal in both output locality and input locality. Our construction is based on the intractability of decoding a random linear code. Previous constructions of PRGs, or even OWFs (cf. [69], Chap. 7), which enjoyed constant input locality and constant output locality at the same time, were based on less standard intractability assumptions.
- We construct a collection of non-interactive commitment schemes, in which the output locality of the commitment function is 4, and its input locality is 3. The security of this scheme also follows from the intractability of decoding a random

---

[1]Our collections are indexed by a public random key. That is, $\{G_z\}_{z \in \{0,1\}^*}$ is a collection of PRGs if for every z the function G_z expands its input and the pair $(z, G_z(x))$ is pseudorandom for random x and z. In a subsequent work [18], it is shown how to upgrade our results and construct a *single* PRG with an optimal locality (as opposed to collection). See Remark 8.2.

linear code. (We can also get a standard non-interactive commitment scheme under the assumption that there exists an explicit binary linear code that has a large minimal distance but is hard to decode.)

- We construct a semantically secure public-key encryption scheme whose encryption algorithm has input locality 3. This scheme is based on the security of the McEliece cryptosystem [110], an assumption which is related to the intractability of decoding a random binary linear code, but is seemingly stronger. Our encryption function also has constant output locality, if the security of the McEliece cryptosystem holds when it is instantiated with some error correcting code whose relative distance is constant.
- We show that MACs, signatures and non-malleable symmetric or public-key encryption schemes cannot be realized by functions whose input locality is constant or, in some cases, even logarithmic in the input length. In fact, we prove that even the weakest versions of these primitives (e.g., one-time secure MACs) cannot be constructed in this model.

Locality-Preserving Reductions We also present new locality-preserving reductions between different cryptographic primitives. Unlike the results discussed above, here we consider *unconditional* reductions that do not rely on unproven assumptions. Specifically, we get new locality-preserving constructions of a one-time symmetric encryption scheme, non-interactive commitment, and a (collection of) PRG from one-to-one OWF. (In fact, in the case of PRG the reduction holds even with more general types of one-way functions.) These reductions preserve both the input locality and the output locality of the underlying primitive up to an additive constant and extend the output locality-preserving reductions from Chaps. 4 and 5.

8.1.2 Our Techniques

Our constructions rely again on the machinery of randomized encoding. In Chap. 4 we showed that the security of most cryptographic primitives is inherited by their randomized encoding. Thus, in order to construct some cryptographic primitive $\mathcal{P}$ in some low complexity class WEAK, one can try to encode functions from a higher complexity class STRONG by functions from WEAK and then take an implementation $f \in$ STRONG of the primitive $\mathcal{P}$, and replace it by its encoding $\hat{f} \in$ WEAK. This paradigm was used in Chaps. 4 and 5 where we showed, for example, that STRONG can be $\mathbf{NC}^1$ (or even $\mathbf{BPP}$) and WEAK can be the class of functions whose output locality is 4.

Adapting this approach to the current setting turns out to be a non-trivial task. While every function can be encoded by an $\mathbf{NC}^0$ function (with sufficiently long output), the power of encodings with constant input locality is extremely limited. It can be shown that such functions *fail* to encode almost all functions including some simple $\mathbf{NC}^0$ functions (see Sect. 8.6).

We overcome this barrier by identifying a simple (but non-trivial) class of functions $\mathscr{C}$, for which we can tailor an encoding with constant input locality. Roughly speaking, a function f is in $\mathscr{C}$ if each of its output bits can be written as a sum of terms over $\mathbb{F}_2$ such that each input variable of f participates in a constant number of *distinct* terms, ranging over all outputs of f. Moreover, if the algebraic degree of these terms is constant, then f can be encoded by a function with constant input locality as well as constant output locality. (In particular, all linear functions over $\mathbb{F}_2$ admit such an encoding.)

By relying on the nice algebraic structure of intractability assumptions related to decoding random linear codes, and using techniques from Chap. 7, we construct PRGs, commitments and public-key encryption schemes in $\mathscr{C}$ whose algebraic degree is constant. Then, we use the new construction to encode these primitives, and obtain implementations whose input locality and output locality are both constant.

Interestingly, unlike previous constructions of randomized encodings, the new encoding does not have a universal simulator nor a universal decoder; that is, one should use different decoders and simulators for different functions in $\mathscr{C}$. This phenomenon is inherent to the setting of constant input locality and is closely related to the fact that MACs cannot be realized in this model. See Sect. 8.6.2 for a discussion.

8.1.3 Previous Work

Unlike the case of $\mathbf{NC}^0$ cryptography, the question of cryptography with constant input locality is relatively unexplored. (A review of the works that address the possibility of $\mathbf{NC}^0$ cryptography appears in Sect. 4.1.1.) However, constructions of primitives with constant input locality are implicitly given in [69, 112]. In particular, Goldreich [69] suggested an approach for constructing OWFs based on expander graphs, an approach whose conjectured security does not follow from any well-known assumption. This general construction can be instantiated by functions with constant output locality and constant input locality. In addition, Mossel et al. [112] constructed (non-cryptographic) ε-biased generators with (non-optimal) constant input and output locality. Also, the construction of linear-stretch PRG in $\mathbf{NC}^0$ of Chap. 7, can give an $\mathbf{NC}^0$ PRG with (large) constant input locality under Assumption 7.1, which is a non-standard assumption taken from [5]. (See Item 2 in Remarks on Theorem 7.3.)

Organization   The rest of this chapter is structured as follows. Following some preliminaries (Sect. 8.2), in Sect. 8.3 we construct a randomized encoding with constant input locality for functions with a "simple" algebraic structure. This construction is used in Sect. 8.4 to derive cryptographic primitives with low locality under coding related assumptions as well as unconditional locality-preserving cryptographic reductions between different primitives. In Sect. 8.5 we prove that MACs and non-malleable encryption schemes cannot be implemented with low input locality. Negative results for randomized encoding with low input locality are discussed in Sect. 8.6.

8.2 Preliminaries

We remind the reader that a polynomial-time function $f : \{0, 1\}^* \to \{0, 1\}^*$ is in the class $\mathbf{Local}_{\text{in}(n)}^{\text{out}(n)}$ if it has an input locality of $\text{in}(n)$ and an output locality of $\text{out}(n)$. (See Sect. 2.4 for more details.)

Locality-Preserving Reductions A *black-box reduction* from the function g to the function f is a polynomial-time algorithm G that computes the function g given oracle access to the function f. We say that such an algorithm G is a *reduction* from a cryptographic primitive $\mathscr{G}$ to a cryptographic primitive $\mathscr{F}$ (or, equivalently, a *construction* of $\mathscr{G}$ from $\mathscr{F}$, or a *transformation* from $\mathscr{F}$ to $\mathscr{G}$) if for any function f which implements $\mathscr{F}$ the function G^f implements the primitive $\mathscr{G}$. When defining $\mathbf{Local}_{\text{in}(n)}^{\text{out}(n)}$ reductions we restrict ourselves to very simple reductions which first query the oracle on some substrings of the original input (in a non-adaptive way) and then perform some $\mathbf{Local}_{\text{in}(n)}^{\text{out}(n)}$ computation. Formally, a reduction from a cryptographic primitive $\mathscr{G}$ to a cryptographic primitive $\mathscr{F}$ is said to be $\mathbf{Local}_{\text{in}(n)}^{\text{out}(n)}$ if it can be written as $G(x) = g(x, f(x^{(1)}), \dots, f(x^{(k)}))$ where the concatenation of $x^{(1)}, \dots, x^{(k)}$ forms a prefix of x, and $g \in \mathbf{Local}_{\text{in}(n)}^{\text{out}(n)}$. A $\mathbf{Local}_{O(1)}^{O(1)}$ reduction G is also called a *locality-preserving reduction* as it preserves the input and output locality of f up to a constant factor.

8.2.1 Cryptographic Primitives

Collections of OWFs and PRGs Standard one-way functions (and their variants) and pseudorandom generators were defined in Chap. 4. In this chapter, we consider collections of PRGs (resp. OWFs, weak OWFs, distributionally OWFs). Let $p(\cdot)$ be a polynomial, and let $\mathscr{G} = \{G_z\}_{z \in \{0,1\}^{p(n)}}$ be a polynomial-time computable collection of functions where $G_z : \{0, 1\}^n \to \{0, 1\}^{s(n)}$. Then $\mathscr{G}$ is a PRG collection (resp. OWF collection, weak OWF collection, distributional OWF collection), if $G'(x, z) = (G_z(x), z)$ is a PRG (resp. OWF, weak OWF, distributional OWF). Note that this definition falls into the category of public-coin collection (as defined in Appendix 4.9). Moreover, in this case the public key is simply a random string.

We will consider a collection of non-interactive commitment schemes. In such a scheme, the sender and the receiver share a common public random key z (that can be selected once and be used in many invocations of the scheme). To commit to a bit b, the sender computes the commitment function $\text{COM}_z(b; r)$ that outputs a commitment σ using the randomness r, and sends the output to the receiver. To open the commitment, the sender sends the randomness r and the committed bit b to the receiver who checks whether the opening is valid by computing the function $\text{REC}_z(\sigma, b, r)$. The scheme should be both (computationally) hiding and (statistically) binding. Hiding requires that $\sigma = \text{COM}_z(b; r)$ keeps b computationally secret. Binding means that, except with negligible probability over the choice of the random public key, there is no commitment string that can be opened in two different ways.

Definition 8.1 (Commitment) A commitment scheme is a pair (COM, REC) where COM is a probabilistic polynomial-time algorithm and REC is a deterministic polynomial-time algorithm. The scheme should satisfy the following conditions:

- **Viability.** For every bit $b \in \{0, 1\}$, $\Pr_{z,r}[\text{REC}_z(\text{COM}_z(b; r), b, r) = \mathsf{reject}] < \text{neg}(|z|)$.
- **Hiding.** $\{(z, \text{COM}_z(0; r))\}_n \overset{c}{\equiv} \{(z, \text{COM}_z(1; r))\}_n$ where $z \leftarrow U_n, r \leftarrow U_{p(n)}$, and the polynomial $p(\cdot)$ is the randomness complexity of COM.
- **Binding.** $\Pr_z[\exists \sigma, r_0, r_1$ such that $\text{REC}_z(\sigma, 0, r_0) = \text{REC}_z(\sigma, 1, r_1) = \mathsf{accept}] < \text{neg}(|z|)$.

In Chap. 4 we defined secure encryption schemes based on the notion of semantic security. A complementary aspect of security is defined via the notion of Non-malleability [57]. Informally, an encryption scheme is said to be non-malleable if it is impossible, given a ciphertext c encrypting a message x, to efficiently generate an encryption c' of a "related" message x' except by copying c. The following definition of Non-malleable private-key encryption is based on [100]. Since this definition is used here for negative results, we allow ourselves to use a simplified version which is weaker than the original definition.

Definition 8.2 (Non-malleable private-key encryption) A *non-malleable private-key encryption scheme* is a triple (G, E, D) of probabilistic polynomial-time algorithms satisfying the following conditions:

- **Viability.** On input 1^n the randomized key generation algorithm, G, outputs a key k. For every $k \in \text{support}(G(1^n))$ and every plaintext x, the algorithms E, D satisfy
$$\Pr\big[D\big(k, E(k, x)\big) \neq x\big] \leq \text{neg}(n),$$
where the probability is taken over the internal randomness of E and D.
- **Non-malleability.** For an adversary A, consider the following experiment which is indexed by n, the size of the key. First a random n-bit key k is selected. Then, A outputs a message space distribution $\mathcal{M}$ which consists of strings of equal length (and is represented by a probabilistic polynomial-sized circuit), and a binary relation R (which is also represented by a polynomial-sized circuit). Next, two random strings $x, \tilde{x}$ are chosen from $\mathcal{M}$, and the ciphertext $c = E(k, x)$ is given to A. Finally, A outputs a ciphertext $c' \neq c$. The advantage of A is defined to be
$$\varepsilon_A(n) = \big|\Pr\big[\big(D(k, c'), x\big) \in R\big] - \Pr\big[\big(D(k, c'), \tilde{x}\big) \in R\big]\big|.$$

The scheme is *non-malleable* if for every (non-uniform) efficient adversary A the advantage $\varepsilon_A(n)$ of A is negligible in n.

8.2.2 Extractors

We will also need the definition of extractors. Recall that the min-entropy of a random variable X is defined as $H_\infty(X) \stackrel{\text{def}}{=} \min_x \log(\frac{1}{\Pr[X=x]})$.

Definition 8.3 (Extractor) A function $\text{Ext}: \{0, 1\}^n \times \{0, 1\}^d \to \{0, 1\}^t$ is a (k, ε)-extractor if for every distribution X on $\{0, 1\}^n$ with $H_\infty(X) \geq k$ the distribution $\text{Ext}(X, U_d)$ is ε-close to the uniform distribution over $\{0, 1\}^t$.

An important construction of extractors is based on pairwise independent hashing.

Lemma 8.1 (Leftover hashing lemma [86]) *Let* $H = \{h_z\}$ *be a family of pairwise independent hash functions that map n-bit strings to t-bit strings. Then, the function* $\text{Ext}(x, z) = (h_z(x), z)$ *is a* (k, ε) *extractor where* $\varepsilon = 2^{-(k-t)/2}$.

We note that pairwise independent hash functions can be defined by the mapping $h_{A,b}(x) = Ax + b$ where A is a $t \times n$ binary matrix, b is a t-bit vector and arithmetic is over $\mathbb{F}_2$.

8.3 Randomized Encoding with Constant Input Locality

In this section we will show that functions with a "simple" algebraic structure (and in particular *linear* functions over $\mathbb{F}_2$) can be encoded by functions with constant input locality.

8.3.1 Key Lemmas

We begin with the following construction that shows how to reduce the input locality of a function which is represented as a sum of functions.

Construction 8.1 (Basic input locality construction) *Let*

$$f(x) = \big(a(x) + b_1(x), a(x) + b_2(x), \ldots, a(x) + b_k(x), c_1(x), \ldots, c_l(x)\big),$$

where $f : \mathbb{F}_2^n \to \mathbb{F}_2^{k+l}$ *and* $a, b_1, \ldots, b_k, c_1, \ldots, c_l : \mathbb{F}_2^n \to \mathbb{F}_2$. *The encoding* $\hat{f} : \mathbb{F}_2^{n+k} \to \mathbb{F}_2^{2k+l}$ *is defined by:*

$$\hat{f}\big(x, (r_1, \ldots, r_k)\big) \stackrel{\text{def}}{=} \big(r_1 + b_1(x), r_2 + b_2(x), \ldots, r_k + b_k(x),$$
$$a(x) - r_1, r_1 - r_2, \ldots, r_{k-1} - r_k,$$
$$c_1(x), \ldots, c_l(x)\big).$$

We refer to a as the pivot *of the construction.*

Note that after the transformation the pivot function $a(x)$ appears only once and therefore the locality of the input variables that appear in a is reduced. In addition, the locality of all the other original input variables does not increase. For example, applying the locality construction to the function $f(x) = (x_1x_2 + x_2, x_1x_2 + x_2x_3, x_1x_2 + x_3, x_3)$ with x_1x_2 as a pivot results in the encoding $\hat{f}(x, r) = (r_1 + x_2, r_2 + x_2x_3, r_3 + x_3, x_1x_2 - r_1, r_1 - r_2, r_2 - r_3, x_3)$. Hence, in this case it reduces the locality of x_1 from 3 to 1.

Lemma 8.2 (Input locality lemma) *Let f and $\hat{f}$ be as in Construction* 8.1. *Then, $\hat{f}$ is a perfect randomized encoding of f.*

Proof The encoding $\hat{f}$ is stretch-preserving since the number of random inputs equals the number of additional outputs (i.e., k). Moreover, given a string $\hat{y} = \hat{f}(x, r)$ we can decode the value of $f(x)$ as follows: To recover $a(x) + b_i(x)$, compute the sum $y_i + y_{k+1} + y_{k+2} + \cdots + y_{k+i}$; to compute $c_i(x)$, simply take y_{2k+i}. This decoder never errs.

Fix some $x \in \{0, 1\}^n$. Let $y = f(x)$ and let $\hat{y}$ denote the distribution $\hat{f}(x, U_k)$. To prove perfect privacy, note that: (1) the last l bits of $\hat{y}$ are fixed and equal to $y_{[k+1\ldots k+l]}$; (2) the first k bits of $\hat{y}$ are independently uniformly distributed; (3) the remaining bits of $\hat{y}$ are uniquely determined by y and $\hat{y}_1, \ldots, \hat{y}_k$. To see (3), observe that, by the definition of $\hat{f}$, we have $\hat{y}_{k+1} = y_1 - \hat{y}_1$; and for every $1 < i \le k$, we also have $\hat{y}_{k+i} = y_i - \hat{y}_i - \sum_{j=1}^{i-1} \hat{y}_{k+j}$.

Hence, define a perfect simulator as follows. Given $y \in \{0, 1\}^{k+l}$, the simulator S chooses a random string r of length k, and outputs $(r, s, y_{[k+1\ldots k+l]})$, where $s_1 = y_1 - r_1$ and $s_i = y_i - r_i - \sum_{j=0}^{i-1} s_j$ for $1 < i \le k$. This simulator is also balanced as each of its outputs is a linear function that contains a fresh random bit. (Namely, the output bit $S(y; r)_i$ depends on: (1) r_i if $1 \le i \le k$; or (2) y_{i-k} if $k + 1 \le i \le 2k + l$.) $\square$

We will also need the following simple transformation.

Lemma 8.3 *Let $f : \mathbb{F}_2^n \to \mathbb{F}_2^{k+l+1}$ be a function of the form*

$$f(x) = \big(a(x), a(x) + b_1(x), a(x) + b_2(x), \ldots, a(x) + b_k(x), c_1(x), \ldots, c_l(x)\big),$$

where $a, b_1, \ldots, b_k, c_1, \ldots, c_l : \mathbb{F}_2^n \to \mathbb{F}_2$. Then the function

$$\hat{f}(x) = \big(a(x), b_1(x), b_2(x), \ldots, b_k(x), c_1(x), \ldots, c_l(x)\big)$$

is a perfect (deterministic) encoding of f. We refer to a as the pivot *of the construction.*

Proof The encoding $\hat{f}$ is stretch-preserving since we did not add any additional outputs and did not use randomness at all. Moreover, there exists a fixed matrix $M \in \mathbb{F}_2^{(k+l+1) \times (k+l+1)}$ of full rank such that $f(x) = M \cdot \hat{f}(x)$ for every x. Hence, the encoding is perfectly private, perfectly correct and balanced. $\square$

Again, after the transformation the locality of the input variables that appear in the pivot a is reduced, while the locality of all the other original input variables does not increase.

8.3.2 Main Results

In the following it will often be useful to take an algebraic view of functions over bit-strings, specifying such functions using an *additive representation* over $\mathbb{F}_2$.

Definition 8.4 (Additive representation) An additive representation of a function $f : \mathbb{F}_2^n \to \mathbb{F}_2^l$ is a representation in which each output bit is written as a sum (over $\mathbb{F}_2$) of functions of the input x. That is, each output bit f_i can be written as $f_i(x) = \sum_{a \in T_i} a(x)$, where T_i is a set of boolean functions over n variables. We specify such an additive representation by an l-tuple $(T_1, \ldots, T_l)$ where T_i is a set of boolean functions $a : \mathbb{F}_2^n \to \mathbb{F}_2$. We assume, without loss of generality, that none of the T_i's contains the constant functions 0 or 1.

For example, any function f whose algebraic degree over $\mathbb{F}_2$ is d admits an additive representation in which each a is a product of at most d input variables.

The following measures are defined with respect to a given additive representation of f.

Definition 8.5 (Multiplicity and rank) For a function $a : \mathbb{F}_2^n \to \mathbb{F}_2$, define the *multiplicity* of a to be the number of T_i's in which a appears, i.e., $\#a = |\{T_i \mid a \in T_i\}|$. Given an additive representation of f, we define the *rank* of a variable x_j to be the number of different boolean functions a which depend on x_j and appear in some T_i. That is,

$$\mathrm{rank}(x_j) = |\{a : \mathbb{F}_2^n \to \mathbb{F}_2 \mid a \text{ depends on } x_j, a \in T_1 \cup \cdots \cup T_l\}|.$$

Theorem 8.1 Let $f : \mathbb{F}_2^n \to \mathbb{F}_2^l$ be a function, and fix some additive representation $(T_1, \ldots, T_l)$ for f. Then f can be perfectly encoded by a function $\hat{f} : \mathbb{F}_2^n \times \mathbb{F}_2^m \to \mathbb{F}_2^s$ such that the following hold:

1. The input locality of every x_j in $\hat{f}$ is at most $\mathrm{rank}(x_j)$, and the input locality of the random inputs r_i of $\hat{f}$ is at most 3.
2. The output locality of $\hat{f}$ is bounded from above by the output locality of f.
3. The randomness complexity of $\hat{f}$ is at most $\sum_{a \in T} \#a$, where $T = \bigcup_{i=1}^l T_i$.

Proof We will use the following convention. The additive representation of a function $\hat{g}$ resulting from applying Construction 8.1 or Lemma 8.3 to a function g is the (natural) representation induced by the original additive representation of g. Let $T = \bigcup_{i=1}^l T_i$ where $(T_1, \ldots, T_l)$ is the original additive representation of f. We construct $\hat{f}$ iteratively via the following process.

- Let $f^{(0)} \leftarrow f$ and $i \leftarrow 0$.
- For all $a \in T$
  - **if** one of the output bits of $f^{(i)}$ is equal to a **then** apply Lemma 8.3 to $f^{(i)}$ with a as a pivot.
  - **elseif** the multiplicity of a in $f^{(i)}$ is greater than 1 **then** apply Construction 8.1 to $f^{(i)}$ with a as a pivot.
  - Record the resulting encoding in $f^{(i+1)}$ and let $i \leftarrow i + 1$.
- Output $\hat{f} \leftarrow f^{(i)}$.

By Lemmas 8.2 and 8.3, the function $f^{(i)}$ perfectly encodes the function $f^{(i-1)}$, hence by the composition property of randomized encodings (Lemma 3.3) the final function $\hat{f}$ perfectly encodes f.

The first item of the theorem follows from the following observations: (1) In each iteration the input locality and the rank of each original variable x_j do not increase. (2) The multiplicity in $\hat{f}$ of every function a that depends on some original input variable x_j is 1. (3) The input locality of the random inputs which are introduced by the locality construction is at most 3.

To prove the second item of the theorem it suffices to show that in each iteration the output locality is not increased. Indeed, Construction 8.1 does not increase the output locality as long as the pivot does not appear as an output bit. Moreover, in the latter case instead of using Construction 8.1 we apply Lemma 8.3 which does not increase the output locality at all.

Finally, the last item follows by noting that the randomness complexity of Construction 8.1 is equal to the multiplicity of the pivot a.   □

Remarks on Theorem 8.1

1. By Theorem 8.1, every linear function admits an encoding of constant input locality, since each output bit can be written as a sum of degree 1 monomials. More generally, every function f whose canonic representation as a sum of monomials (i.e., each output bit is written as a sum of monomials) includes a constant number of monomials per input bit can be encoded by a function of constant input locality.

2. Interestingly, Construction 8.1 does not provide a universal encoding for any natural class of functions (e.g., the class of linear functions mapping n bits into l bits). This is contrasted with previous constructions of randomized encoding with constant output locality (cf. Sect. 4.2). In fact, in Sect. 8.6.1 we prove that there is no universal encoding with constant input locality for the class of linear function $L : \mathbb{F}_2^n \to \mathbb{F}_2$.

3. When Theorem 8.1 is applied to a function family $f : \{0, 1\}^n \to \{0, 1\}^{l(n)}$ then the resulting encoding is uniform whenever the additive representation $(T_1, \dots, T_l)$ is polynomial-time computable.

4. In Sect. 8.6.1, we show that Theorem 8.1 is tight in the sense that for each integer $i > 0$ we can construct a function f in which the rank of x_1 is i, and in every encoding $\hat{f}$ of f the input locality of x_1 is at least i.

In some cases we can combine Theorem 8.1 and the output-locality construction (Construction 4.1) to derive an encoding which enjoys low input locality and output locality at the same time. In particular, we will use the following lemma which is a refinement of Lemma 4.2.

Lemma 8.4 (Implicit in Chap. 4) *Let $f : \mathbb{F}_2^n \to \mathbb{F}_2^l$. Fix some additive representation T for f in which each output bit is written as a sum of monomials of degree (at most) d. Then, we can perfectly encode f by a function $\hat{f}$ with an additive representation $\hat{T}$ such that:*

- $\hat{f} \in \mathbf{Local}^{\max(d+1,3)}$.
- *The rank of every original variable x_i in $\hat{f}$ (with respect to $\hat{T}$) is equal to the rank of x_i in f (with respect to T).*
- *The new variables introduced by $\hat{f}$ appear only in monomials of degree 1; hence their rank is 1.*

By combining Lemma 8.4 with Theorem 8.1 we get:

Corollary 8.1 *Let $f : \mathbb{F}_2^n \to \mathbb{F}_2^l$ be a function. Fix some additive representation for f in which each output bit is written as a sum of monomials of degree (at most) d and the rank of each variable is at most ρ. Then, f can be perfectly encoded by a function $\hat{f}$ of input locality $\max(\rho, 3)$ and output locality $\max(d+1, 3)$. Moreover, the resulting encoding is uniform whenever the additive representation is polynomial-time computable.*

Proof Apply Lemma 8.4 to perfectly encode f by a function $f' \in \mathbf{Local}^{\max(d+1,3)}$ without increasing the rank of the input variables of f. Next, apply Theorem 8.1 to perfectly encode f' by a function $\hat{f} \in \mathbf{Local}_{\max(\rho,3)}^{\max(d+1,3)}$. By the composition property of randomized encodings (Lemma 3.3), the resulting function $\hat{f}$ perfectly encodes f. Finally, the proofs of Theorem 8.1 and Lemma 8.4 both allow us to efficiently transform an additive representation of the function f into an encoding $\hat{f}$ in $\mathbf{Local}_{\max(\rho,3)}^{\max(d+1,3)}$. Hence, the uniformity of f is inherited by $\hat{f}$. $\square$

It can be shown that if each output bit of f can be written as the sum of at most t degree-d monomials, then the randomness and output complexity of the above encoding is at most $O(tl)$. We also remark that Theorem 8.1, Lemma 8.4, and Corollary 8.1 generalize to any finite field $\mathbb{F}$.

Remark 8.1 By Corollary 8.1 any linear (or affine) function $L : \mathbb{F}_2^n \to \mathbb{F}_2^l$ can be encoded by a function $\hat{L}(x, r) \in \mathbf{Local}_3^3$. Moreover, a closer look at Theorem 8.1 and Construction 4.1 shows that $\hat{L}$ has the following additional properties: (1) the input locality of the x's is 1; (2) the outputs that depend on the x's have (output) locality 2; and (3) the randomness and output complexity of $\hat{L}$ is $O(n \cdot \min(n, l))$.

8.4 Primitives with Constant Input Locality and Output Locality

8.4.1 Main Assumption: Intractability of Decoding Random Linear Code

Our positive results are based on the intractability of decoding a random linear code. In the following we introduce and formalize this assumption.

An (m, n, δ) *binary linear code* is an n-dimensional linear subspace of $\mathbb{F}_2^m$ in which the Hamming distance between each two distinct vectors (codewords) is at least δm. We refer to the ratio n/m as the *rate* of the code and to δ as its (relative) *distance*. Such a code can be defined by an $m \times n$ *generator matrix* whose columns span the space of codewords. It follows from the Gilbert-Varshamov bound that whenever $n/m < 1 - \mathrm{H}_2(\delta) - \varepsilon$, almost all $m \times n$ generator matrices form (m, n, δ)-linear codes. Formally, we have the following.

Fact 8.1 ([135]) *Let $0 < \delta < 1/2$ and $\varepsilon > 0$. Let $n/m \le 1 - \mathrm{H}_2(\delta) - \varepsilon$. Then, a randomly chosen $m \times n$ generator matrix generates an (m, n, δ) code with probability $1 - 2^{-(\varepsilon/2)m}$.*

A proof of the above version of the Gilbert-Varshamov bound can be found in Lecture 5 of [132]. For code length parameter $m = m(n)$, and noise parameter $\mu = \mu(n)$, we will consider the following "decoding game". Pick a random $m \times n$ matrix C representing a linear code, and a random information word x. Encode x with C and transmit the resulting codeword $y = Cx$ over a binary symmetric channel in which every bit is flipped with probability μ. Output the noisy codeword $\tilde{y}$ along with the code's description C. The adversary's task is to find the information word x. We say that the above game is intractable if every polynomial-time adversary wins in the above game with no more than negligible probability in n.

Definition 8.6 Let $m(n) \le \mathrm{poly}(n)$ be a code length parameter, and $0 < \mu(n) < 1/2$ be a noise parameter. The problem $\mathrm{CODE}(m, \mu)$ is defined as follows:

- **Input**: $(C, Cx + e)$, where C is an $m(n) \times n$ random binary generator matrix, $x \leftarrow U_n$, and $e \in \{0, 1\}^m$ is a random error vector in which each entry is chosen to be 1 with probability μ (independently of other entries), and arithmetic is over $\mathbb{F}_2$.
- **Output**: x.

We say that $\mathrm{CODE}(m, \mu)$ is *intractable* if every (non-uniform) polynomial-time adversary A solves the problem with probability negligible in n.

We note that $\mathrm{CODE}(m, \mu)$ becomes harder when m is decreased and μ is increased, as we can always add noise or ignore the suffix of the noisy codeword. Formally, we have the following.

Proposition 8.1 *Let $m'(n) \leq m(n)$ and $0 < \mu(n) \leq \mu'(n) < 1/2$ for every n. Then, if* CODE(m, μ) *is intractable, so is* CODE(m', μ').

Proof Fix n and let $m' = m'(n), m = m(n), \mu = \mu(n)$ and $\mu' = \mu'(n)$. We reduce the problem CODE(m, μ) to CODE(m', μ') as follows. Given an input (C, y) for CODE(m, μ) (i.e., C is an $m \times n$ binary matrix and y is an m-bit vector), we construct the pair (C', y') by letting C' denote the $m' \times n$ binary matrix that contains the first m' rows of C, and $y' \in \{0, 1\}^{m'}$ be the vector resulting by taking the first m' entries of y and adding (over $\mathbb{F}_2$) a random vector $r \in \{0, 1\}^{m'}$ in which each entry is chosen to be 1 (independently of other entries) with probability $(\mu' - \mu)/(1 - 2\mu)$.

Suppose that (C, y) is drawn from the input distribution of CODE(m, μ), that is, C is random matrix and $y = Cx + e$ where $x \leftarrow U_n$, and $e \in \{0, 1\}^m$ is a random error vector of noise rate μ. Then, the pair (C', y') can be written as $(C', C'x + e')$ where $e' = e + r$ is a random noise vector of rate

$$\mu \cdot \left(1 - \frac{\mu' - \mu}{1 - 2\mu}\right) + (1 - \mu)\frac{\mu' - \mu}{1 - 2\mu} = \mu + \frac{(1 - 2\mu)(\mu' - \mu)}{1 - 2\mu} = \mu'.$$

Hence, given an algorithm A that solves CODE(m', μ'), we can find the information word x by running A on (C', y'). $\square$

The hardness of CODE(m, μ) is well studied [32, 33, 62, 88, 97, 102, 109]. It can be also formulated as the problem of learning parity with noise, and it is known to be NP-complete in the worst-case [31]. It is widely believed that the problem is hard for every fixed μ and every $m(n) \in O(n)$, or even $m(n) \in \text{poly}(n)$. Similar assumptions were put forward in [32, 70, 76, 88, 97, 99, 101]. The plausibility of such an assumption is supported by the fact that a successful attack would imply a major breakthrough in coding theory. We mention that the best known algorithm for CODE(m, μ), due to Blum et al. [33], runs in time $2^{O(n/\log n)}$ and requires m to be $2^{O(n/\log n)}$. Lyubashevsky [109] showed how to reduce m to be only superlinear, i.e., $n^{1+\alpha}$, at the cost of increasing the running time to $2^{O(n/\log \log n)}$. When $m = O(n)$ (and μ is constant), the problem is only known to be solved in *exponential* time.

Our Parameters Typically, we let $m(n) \in O(n)$ and μ be a constant such that $n/m(n) < 1 - H_2(\mu + \varepsilon)$ where $\varepsilon > 0$ is a constant. Hence, by Fact 8.1, the random code C is, with overwhelming probability, an $(m, n, \mu + \varepsilon)$ code. Note that, except with negligible probability, the noise vector flips less than $\mu + \varepsilon$ of the bits of y. In this case, the fact that the noise is random (rather than adversarial) guarantees, by Shannon's coding theorem (for random linear codes), that x will be unique with overwhelming probability. That is, roughly speaking, we assume that it is intractable to correct μn *random* errors in a random linear code of relative distance $\mu + \varepsilon > \mu$ and some (fixed) constant rate.

Pseudorandomness We now show that distinguishing the distribution $(C, Cx + e)$ from the uniform distribution reduces to decoding x. A similar lemma was proved by Blum et al. [32, Theorem 13]. However, their version does not preserve the length of the codewords. Namely, they show that the hardness of decoding random linear code with codewords of length $m(n)$ implies the pseudorandomness of the distribution $(C, Cx + e)$ in which the length of the codewords is *polynomially smaller* than $m(n)$.

Lemma 8.5 *Let $m(n)$ be a code length parameter, and $\mu(n)$ be a noise parameter. If $\text{CODE}(m, \mu)$ is intractable then the distribution $(C, Cx + e)$ is pseudorandom, where $C \leftarrow U_{m(n) \times n}$, $x \leftarrow U_n$, and $e \in \{0, 1\}^{m(n)}$ is a random error vector of noise rate μ.*

Proof Assume that $\text{CODE}(m, \mu)$ is intractable. Then, by the Goldreich-Levin hardcore bit theorem [77], given $(C, Cx + e)$ and a random n-bit vector r, an efficient adversary cannot compute $\langle r, x \rangle$ with probability greater than $\frac{1}{2} + \text{neg}(n)$. Assume, towards a contradiction, that there exists an efficient distinguisher $A = \{A_n\}$ and a polynomial $p(\cdot)$ such that

$$\Pr\big[A_n(C, Cx + e) = 0\big] - \Pr\big[A_n(U_{m(n) \times n}, U_m) = 0\big] > 1/p(n),$$

for infinitely many n's. We will use A_n to construct an efficient adversary A'_n that breaks the security of the Goldreich-Levin hardcore bit. Given $(C, y = Cx + e)$ and a random n-bit vector r, the adversary A'_n chooses a random m-bit vector s and computes a new $m(n) \times n$ binary matrix $C' \stackrel{\text{def}}{=} C - s \cdot r^T$, where r^T denotes the transpose of r. Now A'_n applies the distinguisher A_n to (C', y) and outputs his answer. Before we analyze the success probability of A'_n we need two observations: (1) the matrix C' is a random $m(n) \times n$ binary matrix; and (2) $y = Cx + e = C'x + s \cdot r^T \cdot x + e = C'x + s \cdot \langle r, x \rangle + e$. Hence, when $\langle r, x \rangle = 0$ it holds that $(C', y) = (C', C'x + e)$, and when $\langle r, x \rangle = 1$ we have $(C', y) = (C', C'x + e + s) \equiv (C', U_m)$, where U_m is independent of C'. Therefore we have

$$\Pr\big[A'_n(C, Cx + e, r) = \langle x, r \rangle\big]$$
$$= \Pr\big[A'_n(C, Cx + e, r) = 0 | \langle x, r \rangle = 0\big] \cdot \Pr\big[\langle x, r \rangle = 0\big]$$
$$+ \Pr\big[A'_n(C, Cx + e, r) = 1 | \langle x, r \rangle = 1\big] \cdot \Pr\big[\langle x, r \rangle = 1\big]$$
$$= \frac{1}{2} \cdot \big(\Pr\big[A_n(C', C'x + e) = 0\big] + 1 - \Pr\big[A_n(C', U_m) = 0\big]\big)$$
$$\geq \frac{1}{2} + \frac{1}{2p(n)},$$

where the last inequality holds for infinitely many n's. Thus, we derive a contradiction to the security of the GL-hardcore bit.   □

8.4.2 Pseudorandom Generator in $\mathbf{Local}_3^3$

A pseudorandom generator (PRG) is an efficiently computable function G which expands its input and its output distribution $G(U_n)$ is pseudorandom. An efficiently computable collection of functions $\{G_z\}_{z \in \{0,1\}^*}$ is a PRG collection if for every z, the function G_z expands its input and the pair $(z, G_z(x))$ is pseudorandom for random x and z. (See Sect. 8.2 for formal definitions.) We show that pseudorandom generators (and therefore also one-way functions and one-time symmetric encryption schemes) can be realized by $\mathbf{Local}_{O(1)}^{O(1)}$ functions. Specifically, we get a PRG in $\mathbf{Local}_3^3$. Recall that, by the tractability of 2-SAT, it is impossible to construct a PRG (and even OWF) in $\mathbf{Local}^2$ [50, 69]. In Sect. 8.5.4 we also prove that there is no PRG in $\mathbf{Local}_2$. Hence, our PRG has optimal input locality as well as optimal output locality.

We rely on the following assumption.

Intractability Assumption 8.1   *The problem* $\mathrm{CODE}(6n, 1/4)$ *is intractable.*

Note that the code considered here is of rate $n/m = 1/6$ which is strictly smaller than $1 - H_2(1/4)$. Therefore, except with negligible probability, its relative distance is larger than $1/4$. Hence, the above assumption roughly says that it is intractable to correct $n/4$ random errors in a random linear code of relative distance $1/4 + \varepsilon$, for some constant $\varepsilon > 0$.

Let $m(n) = 6n$. Let $C \leftarrow U_{m(n) \times n}$, $x \leftarrow U_n$ and $e \in \{0, 1\}^m$ be a random error vector of rate $1/4$, that is, each of the entries of e is 1 with probability $1/4$ (independently of the other entries). By Lemma 8.5, the distribution $(C, Cx + e)$ is pseudorandom under the above assumption. Since the noise rate is $1/4$, it is natural to sample the noise distribution e by using $2m$ random bits $r_1, \ldots, r_{2m}$ and letting the i-th bit of e be the product of two fresh random bits, i.e., $e_i = r_{2i-1} \cdot r_{2i}$. We can now define the mapping $f(C, x, r) = (C, Cx + e(r))$ where $e(r) = (r_{2i-1} \cdot r_{2i})_{i=1}^m$. The output distribution of f is pseudorandom, however, f is not a PRG since it does not expand its input. In Chap. 7, we showed how to bypass this problem by applying a randomness extractor (see Definition 8.3). Namely, the following function was shown to be a PRG: $G(C, x, r, s) = (C, Cx + e(r), \mathrm{Ext}(r, s))$. Although the setting of parameters in Chap. 7 is different than ours, a similar solution works here as well. We rely on the leftover hashing lemma (Lemma 8.1) and base our extractor on a family of pairwise independent hash functions (which is realized by the mapping $x \mapsto Ax + b$ where A is a random matrix and b is a random vector).[2]

Construction 8.2   *Let* $m = 6n$ *and let* $t = \lceil 7.1 \cdot n \rceil$. *Define the function*

$$G(x, C, r, A, b) \stackrel{\mathrm{def}}{=} (C, Cx + e(r), Ar + b, A, b),$$

where $x \in \{0, 1\}^n$, $C \in \{0, 1\}^{m \times n}$, $r \in \{0, 1\}^{2m}$, $A \in \{0, 1\}^{t \times 2m}$, *and* $b \in \{0, 1\}^t$.

---

[2]We remark that in Chap. 7 one had to rely on a specially made extractor in order to maintain the large stretch of the PRG. In particular, the leftover hashing lemma could not be used there.

Theorem 8.2 *Under Assumption 8.1, the function G defined in Construction 8.2 is a PRG.*

Before proving Theorem 8.2, we need the following claim (which is similar to Lemma 7.3).

Claim 8.1 *Let $[r|e(r)]$ denote the distribution of r given the outcome of $e(r)$. Then,*

$$\Pr_{r \leftarrow U_{2m}} \left[H_\infty([r|e(r)]) \geq 1.17m \right] \geq 1 - \exp(-\Omega(m)).$$

Proof We view $e(r)$ as a sequence of m independent Bernoulli trials, each with a probability 0.25 of success. Recall that r is composed of m pairs of bits, and that the i-th bit of $e(r)$ is 1 if and only if r_{2i-1} and r_{2i} are both equal to 1. Hence, whenever $e(r)_i = 0$, the pair (r_{2i-1}, r_{2i}) is uniformly distributed over the set $\{00, 01, 10\}$. Consider the case in which at most $0.26m$ components of $e(r)$ are ones. By a Chernoff bound, the probability of this event is at least $1 - \exp(-\Omega(m))$. In this case, r is uniformly distributed over a set of size at least $3^{0.74m}$. Hence, conditioning on the event that at most $0.26m$ components of $e(r)$ are ones, the min-entropy of $[r|e(r)]$ is at least $0.74m \log(3) > 1.17m$. $\square$

We can now prove Theorem 8.2.

Proof of Theorem 8.2 Let $m = 6n$ and $t = 7.01n$. First we show that G expands its input. Indeed, the difference between the output length and the input length is: $m + t - (n + 2m) = 0.01n > 0$.

Let $x \leftarrow U_n$, $C \leftarrow U_{m \cdot n}$, $r \leftarrow U_{2m}$, $A \leftarrow U_{t \cdot 2m}$ and $b \leftarrow U_t$. We prove that the distribution $G(x, C, r, A, b)$ is pseudorandom. By Lemma 8.1, Fact 2.4 and Claim 8.1, we have that

$$\left\| (e(r), Ar + b, A, b) - (e(r), U_{t+2tm+m}, A, b) \right\| \leq 2^{-(1.17m-t)/2} + \exp(-\Omega(m))$$

$$= 2^{-0.005n} + \exp(-\Omega(n))$$

$$\leq \exp(-\Omega(n)).$$

Hence, by Fact 2.3 and Assumption 8.1, we have

$$(C, Cx + e(r), Ar + b, A, b) \overset{s}{\equiv} (C, Cx + e(r), U_{t+2tm+m}) \overset{c}{\equiv} U_{mn+m+t+2tm+m},$$

which completes the proof. $\square$

From now on, we fix the parameters m, t according to Construction 8.2. We can redefine the above construction as a collection of PRGs by letting C, A, b be the keys of the collection. Namely,

$$G_{C,A,v}(x, r) = (Cx + e(r), Ar + v). \tag{8.1}$$

We can now prove the main theorem of this section.

Theorem 8.3 *Under Assumption* 8.1, *there exists a collection of pseudorandom generators* $\{G_z\}_{z \in \{0,1\}^{p(n)}}$ *in* **Local**$_3^3$. *Namely, for every* $z \in \{0, 1\}^{p(n)}$, *it holds that* $G_z \in$ **Local**$_3^3$.

Proof Fix C, A, b and write each output bit of $G_{C,A,b}(x, r)$ as a sum of monomials. Note that in this case, each variable x_i appears only in degree-1 monomials, and each variable r_i appears only in the monomial $r_{2i-1}r_{2i}$ and also in degree-1 monomials. Hence, the rank of each variable is at most 2. Moreover, the (algebraic) degree of each output bit of $G_{C,A,b}$ is at most 2. Therefore, by Corollary 8.1, we can perfectly encode the function $G_{C,A,b}$ by a function $\hat{G}_{C,A,b}$ in **Local**$_3^3$. In Chap. 4 it was shown that a uniform perfect encoding of a PRG is also a PRG (see Lemma 4.7). Thus, we get a collection of PRGs in **Local**$_3^3$.   □

Since the encoding $\hat{G}_{C,A,v}$ has $N = \Theta(n^2)$ inputs and $N + \Theta(n)$ outputs we get a pseudorandom generator whose stretch is only *sublinear* in the input length. We mention that, by relying on the results of Chap. 7, one can obtain a PRG with linear stretch and (large) constant input and output locality. However, the security of this construction is based on a non-standard intractability assumption taken from [5].

Remark 8.2 (**Single PRG with optimal locality**)   Theorem 8.2 gives a PRG G of degree-2. By applying the output locality reduction of Chap. 4 (see also Lemma 8.4), we can encode G by a function $\hat{G}$ in **Local**^{3} and get a *single* PRG (rather than a collection of PRGs) with optimal output locality. In a subsequent work [18], it is shown how to improve this result and obtain a single PRG in **Local**$_3^3$ under the same intractability assumption.

8.4.3 Symmetric Encryption

We can rely on Theorem 8.3 to obtain a (collection of) one-time semantically secure symmetric encryption scheme (E_z, D_z) with low input and output locality (whose key is shorter than the message). Specifically, for a (public) collection key z, a private key k, a plaintext x, and a ciphertext c, we define the scheme $(E_z(k, x) = G_z(k) + x, D_z(k, c) = G_z(k) + c$. It is not hard to prove the security of the scheme assuming that G_z is a collection of PRGs. We can instantiate this scheme with the PRG of Theorem 8.3 and obtain an encryption scheme whose both encryption algorithm and decryption algorithm are in **Local**$_3^4$. However this scheme is restricted to encrypt messages whose length is only slightly larger than the key length (as the stretch of G is only sublinear).

We can remove this limitation and also obtain an encryption algorithm in **Local**$_3^3$, at the expense of increasing the complexity of the decryption algorithm. The idea is to use a variant of the aforementioned construction. In particular, recall that

Construction 5.1 uses a PRG (with a one-bit stretch) to obtain a one-time semantically secure symmetric encryption (E, D) that allows us to encrypt an arbitrary polynomially long message with a short key. The encryption algorithm is de-
fined as follows: $E(k, x, (s_1, \ldots, s_{\ell-1})) \overset{\text{def}}{=} (G(k) + s_1, G(s_1) + s_2, \ldots, G(s_{\ell-2}) +$
$s_{l-1}, G(s_{\ell-1}) + x)$, where $k \leftarrow U_n$ is the private key, x is a $(k + \ell)$-bit plaintext and
$s_i \leftarrow U_{n+i}$ serve as the coin tosses of E. If we instantiate this scheme with the PRG
collection G_z of Eq. (8.1), we get a collection of encryption function E_z in which
both the rank and the degree of E_z are at most 2. Hence, by employing Corollary 8.1,
we can encode E_z by a function $\hat{E}_z$ in $\textbf{Local}_3^3$. In Chap. 4 it was shown that in this
case $\hat{E}_z$ forms a one-time semantically secure encryption scheme (together with the
decryption function $\hat{D}_z(k, \hat{c}) = D_z(k, B(\hat{c}))$, where B is the decoding algorithm of
the encoding). Hence, we get a one-time semantically secure symmetric encryption
in $\textbf{Local}_3^3$. However, the decryption is no longer in $\textbf{Local}_{O(1)}^{O(1)}$.

A similar approach can be also used to give multiple message security, at the price
of requiring the encryption and decryption algorithms to maintain a synchronized
state. The results of Sect. 8.4.5 give a direct construction of public-key encryption
(hence also symmetric encryption) with constant input locality under the stronger
assumption that the McEliece cryptosystem is one-way secure. We also mention
that the techniques of [18] achieve similar results (multiple message security and
encryption with constant input and output locality) assuming that $CODE(m, \mu)$ is
intractable for a constant μ (say $1/8$) and every polynomial $m(n)$. The latter seems
to be a weaker (and therefore better) assumption than the security of the McEliece
cryptosystem.

8.4.4 Commitment in Local$_3^4$

We construct a collection of commitment schemes in $\textbf{Local}_3^4$ (i.e., a commitment of
input locality 3 and output locality 4) under the following assumption.

Intractability Assumption 8.2 *There exists a constant c that satisfies $c >$
$\frac{1}{1-H_2(1/4)}$, for which the problem $CODE(\lceil cn \rceil, 1/8)$ is intractable.*

We begin by constructing a commitment scheme COM_z with low rank and low
algebraic degree. Suppose that Assumption 8.2 holds with respect to c (for concreteness we may think of $c = 5.3$). Let $m = m(n) = \lceil cn \rceil$. The public key of our
scheme will be a random $m \times n$ generator matrix C. To commit to a bit b, we
first choose a random information word $x \in \{0, 1\}^n$, hide it by computing $Cx + e$,
where $e \in \{0, 1\}^m$ is a noise vector of rate $1/8$, and then take the exclusive-or of b
with a hardcore bit $\beta(x)$ of the above function. Assuming that $CODE(m, 1/8)$ is
intractable, this commitment hides the committed bit b. To see that the scheme is
binding, recall that by Fact 8.1, the matrix C almost always generates a code whose
relative distance is $1/4 + \varepsilon$, for some constant $\varepsilon > 0$. Suppose that the relative distance of C is indeed $1/4 + \varepsilon$. Then, if e contains no more than $1/8 + \varepsilon/2$ ones,

x is uniquely determined by $Cx + e$. Of course, the sender might try to cheat and open the commitment ambiguously by claiming that the weight of the error vector is larger than $(1/8 + \varepsilon/2) \cdot m$. Hence, we let the receiver verify that the Hamming weight of the noise vector e given to him by the sender in the opening phase is indeed smaller than $(1/8 + \varepsilon/2) \cdot m$. This way, the receiver will always catch a cheating sender (assuming that C is indeed a good code).

Construction 8.3   *Given a constant c that satisfies $c > \frac{1}{1 - H_2(1/4)}$, let $\varepsilon > 0$ be a constant for which $c > \frac{1}{1 - H_2(1/4 + \varepsilon)}$, and let $m = m(n) = \lceil cn \rceil$. We define the following scheme:*

- *Common random key: a random $m \times n$ generator matrix C.*
- *Commitment algorithm: $\text{COM}_C(b; (x, r, s)) = (Cx + e(r), s, b + \langle x, s \rangle)$, where $x, s \leftarrow U_n, r \leftarrow U_{3m}$, and $e(r) = (r_1 r_2 r_3, r_4 r_5 r_6, \ldots, r_{3m-2} r_{3m-1} r_{3m})$.*
- *Receiver algorithm: $\text{REC}_C(\sigma, b, (x, r, s)) = $ accept if and only if $\text{COM}_C(b; (x, r, s)) = \sigma$ and the Hamming weight of the noise vector $e(r)$ is smaller than $(1/8 + \varepsilon/2) \cdot m$.*

Theorem 8.4   *Suppose that Assumption 8.2 holds with respect to the constant c. Then, the scheme defined in Construction 8.3 (instantiated with c) forms a collection of non-interactive commitment schemes.*

Proof (1) Viability: An honest sender will be rejected only if its randomly chosen noise vector $e(r)$ is heavier than $(1/8 + \varepsilon/2) \cdot m$, which, by a Chernoff bound, happens with negligible probability (i.e., $2^{-\Omega(n)}$) as the noise rate is $1/8$.

(2) Hiding: Let C, x, s, r be distributed as in Construction 8.3. Then, by Assumption 8.2 and the Goldreich-Levin theorem [77], we have

$$\left(C, \text{COM}_C\big(0; (x, r, s)\big)\right) \equiv \left(C, \big(Cx + e(r), s, \langle x, s \rangle\big)\right)$$

$$\stackrel{c}{\equiv} \left(C, \big(Cx + e(r), s, U_1\big)\right)$$

$$\equiv \left(C, \big(Cx + e(r), s, 1 + U_1\big)\right)$$

$$\stackrel{c}{\equiv} \left(C, \big(Cx + e(r), s, 1 + \langle x, s \rangle\big)\right)$$

$$\equiv \left(C, \text{COM}_C\big(1; (x, r, s)\big)\right).$$

(3) Binding: Suppose that σ is an ambiguous commitment string. Namely, $\text{REC}_C(\sigma, 0, (x, r, s)) = \text{REC}_C(\sigma, 1, (x', r', s)) = $ accept for some x, r, s, x', r' of appropriate length. Then, there are two distinct code words Cx and Cx' for which $Cx + e(r) = Cx' + e(r')$. Since $e(r)$ and $e(r')$ are of weight smaller than $(1/8 + \varepsilon/2) \cdot m$, we conclude that the relative distance of the code C is smaller than $2 \cdot (1/8 + \varepsilon/2) = 1/4 + \varepsilon$. However, by Fact 8.1, this event happens only with negligible probability (i.e., $2^{-\Omega(m)} = 2^{-\Omega(n)}$) over the choice of C. $\square$

When C is fixed, the rank and algebraic degree of the function COM_C are 2 and 3 (with respect to the natural representation as a sum of monomials). Hence, by

Corollary 8.1, we can encode COM_C by a function $\hat{\text{COM}}_C \in \textbf{Local}_3^4$. By Chap. 4, this encoding is also a commitment scheme. Summarizing, we have the following.

Theorem 8.5 *Under Assumption* 8.2, *there exists a collection of commitment schemes* (COM, REC) *in* $\textbf{Local}_3^4$; *i.e., for every public key* C, *we have* $\text{COM}_C \in \textbf{Local}_3^4$.

We remark that we can get a standard non-interactive commitment (rather than a collection of commitment schemes) by letting C be a generator matrix of some fixed error correcting code whose relative distance is large (i.e., $1/4$ or any other constant) in which decoding is intractable. For example, one might use the dual of a BCH code.

8.4.5 Semantically Secure Public-Key Encryption in $\textbf{Local}_3^{O(1)}$

We construct a semantically secure public-key encryption scheme (PKE) whose encryption algorithm is in $\textbf{Local}_{O(1)}^{O(1)}$ (see Definition 4.8). Our scheme is based on the McEliece cryptosystem [110]. We begin by reviewing the general scheme proposed by McEliece.

- **System parameters.** Let $m(n) : \mathbb{N} \to \mathbb{N}$, where $m(n) > n$, and $\mu(n) : \mathbb{N} \to (0, 1)$. For every $n \in \mathbb{N}$, let $\mathscr{C}_n$ be a set of generating matrices of $(m(n), n, 2(\mu(n) + \varepsilon))$ codes that have a (universal) efficient decoding algorithm D that, given a generating matrix from $\mathscr{C}_n$, can correct up to $(\mu(n) + \varepsilon) \cdot m(n)$ errors, where $\varepsilon > 0$ is some constant.[3] We also assume that there exists an efficient sampling algorithm that samples a generator matrix of a random code from $\mathscr{C}_n$.
- **Key Generation.** Given a security parameter 1^n, use the sampling algorithm to choose a random code from $\mathscr{C}_n$ and let C be its generating matrix. Let $m = m(n)$ and $\mu = \mu(n)$. Choose a random $n \times n$ non-singular matrix S over $\mathbb{F}_2$, and a random $m \times m$ permutation matrix P. Let $C' = P \cdot C \cdot S$ be the public key and P, S, D_C be the private key where D_C is the efficient decoding algorithm of C.
- **Encryption.** To encrypt $x \in \{0, 1\}^n$ compute $c = C'x + e$ where $e \in \{0, 1\}^m$ is an error vector of noise rate μ.
- **Decryption.** To decrypt a ciphertext c, compute $P^{-1}y = P^{-1}(C'x + e) = CSx + P^{-1}e = CSx + e'$ where e' is a vector whose weight equals the weight of e (since P^{-1} is also a permutation matrix). Now, use the decoding algorithm D to recover the information word Sx (i.e., $D(C, CSx + P^{-1}e) = Sx$). Finally, to get x multiply Sx on the left by S^{-1}.

---

[3] In fact, we may allow ε to decrease with n. In such case, we might get a non-negligible decryption error. This can be fixed (without increasing the rank or the degree of the encryption function) by repeating the encryption with independent fresh randomness. Details omitted.

By a Chernoff bound, the weight of the error vector e is, except with negligible probability, smaller than $(\mu + \varepsilon) \cdot m$ and so the decryption algorithm almost never errs. As for the security of the scheme, it is not hard to see that the scheme is *not* semantically secure. (For example, it is easy to verify that a ciphertext c is an encryption of a given plaintext x by checking whether the weight of $c - Cx$ is approximately μn.)

However, the scheme is conjectured to be a one-way cryptosystem; namely, it is widely believed that, for proper choice of parameters, any efficient adversary fails with probability $1 - \text{neg}(n)$ to recover x from $(c = C'x + e, C')$ where x is a random n-bit string. (In other words, the McEliece cryptosystem is considered to be a collection of trapdoor one-way functions which is almost 1-1 with respect to its first argument; i.e., x.)

Suppose that the scheme is indeed one-way with respect to the parameters $m(n)$, $\mu(n)$ and $\mathscr{C}_n$. Then, we can convert it into a semantically secure public-key encryption scheme by extracting a hardcore predicate and XORing it with a 1-bit plaintext b (this transformation is similar to the one used for commitments in the previous section). That is, we encrypt the bit b by the ciphertext $(C'x + e, s, \langle s, x \rangle + b)$ where x, s are random n-bit strings, and e is a noise vector of rate μ. (Again, we use the Goldreich-Levin hardcore predicate [77].) To decrypt the message, we first compute x, by invoking the McEliece decryption algorithm, and then compute $\langle s, x \rangle$ and XOR it with the last entry of the ciphertext. We refer to this scheme as the *modified* McEliece public-key encryption scheme. If the McEliece cryptosystem is indeed one-way, then $\langle s, x \rangle$ is pseudorandom given $(C', C'x + e, s)$, and thus the modified McEliece public-key encryption scheme is semantically secure. Formally,

Lemma 8.6 *If the McEliece cryptosystem is one-way with respect to the parameters $m(n)$, $\mu(n)$ and $\mathscr{C}_n$, then the modified McEliece PKE is semantically secure with respect to the same parameters.*

The proof of this lemma is essentially the same as the proof of Proposition 5.3.14 in [72].

Let $\mu(n) = 2^{-t(n)}$. Then, we can sample the noise vector e by using the function $e(r) = (\prod_{j=1}^{t} r_{t \cdot (i-1)+j})_{i=1}^{m(n)}$ where r is a $t(n) \cdot m(n)$ bit string. In this case, we can write the encryption function of the modified McEliece as $E_{C'}(b, x, r, s) = (C'x + e(r), s, \langle x, s \rangle + b)$.

The rank of each variable of this function is at most 2, and its algebraic degree is at most $t(n)$. Hence, by Corollary 8.1, we can encode it by a function $\hat{E} \in \textbf{Local}_3^{t(n)+1}$, i.e., the output locality of $\hat{E}$ is $t(n) + 1$ and its input locality is 3. In Lemma 4.10 we showed that randomized encoding preserves the security of PKE. Namely, if (G, E, D) is a semantically secure PKE then $(G, \hat{E}, \hat{D})$ is also an encryption scheme where $\hat{E}$ is an encoding of E, $\hat{D}(c) = D(B(c))$ and B is the decoder of the encoding. Hence we have the following.

Theorem 8.6 *If the McEliece cryptosystem is one-way with respect to the parameters $m(n)$, $\mu(n) = 2^{-t(n)}$ and $\mathscr{C}_n$, then there exists a semantically secure PKE whose encryption algorithm is in $\mathbf{Local}_3^{t(n)}$.*

The scheme we construct encrypts a single bit, however we can use concatenation to derive a PKE for messages of arbitrary (polynomial) length without increasing the input and output locality. Theorem 8.6 gives a PKE with constant output locality whenever the noise rate μ is constant. Unfortunately, the binary classical Goppa Codes, which are commonly used with the McEliece scheme [110], are known to have an efficient decoding only for *subconstant* noise rate. Hence, we cannot use them for the purpose of achieving constant output locality and constant input locality simultaneously. Instead, we suggest using algebraic-geometric (AG) codes which generalize the classical Goppa Codes and enjoy an efficient decoding algorithm for constant noise rate. It seems that the use of such codes does not decrease the security of the McEliece cryptosystem [96].

Remark 8.3 Our description of the McEliece cryptosystem assumes that the error-correcting code being used is efficiently decodable from a *constant* noise rate. Specifically, we assume that codes from $\mathscr{C}_n$ can correct up to $(\mu(n) + \varepsilon) \cdot m(n)$ errors, for some *constant* $\varepsilon > 0$. This requirement can be waived. In particular, we may allow $\varepsilon = \varepsilon(n)$ to decrease with n as long as it is larger than, say, $1/\sqrt{m(n)}$. In this case, by a Chernoff bound, the decryption algorithm errs with probability at most $\exp(-2\varepsilon^2 m) < 1/2$. This error can be decreased to negligible by repeating the encryption (of the modified McEliece scheme) $\Omega(n)$ times with independent fresh randomness, and by taking the majority while decrypting. Note that this transformation does not increase the rank or the degree of the encryption function.

Remark 8.4 Recall that in the standard definition of semantic security the adversary's goal is to find two messages x and x', whose encryptions can be distinguished. In particular, x and x' should be chosen before the adversary sees the public key. One may consider a stronger variant of semantic security in which the choice of the pair x and x' may depend on the public key e. It is not hard to show that Lemma 8.6 holds also with respect to this notion of security. (See Sect. 5.4.2 of [72].) Furthermore, randomized encoding preserves semantic-security under key-dependent attacks (Chap. 4), and therefore Theorem 8.6 extends to this setting as well.

8.4.6 Locality-Preserving Reductions Between Different Primitives

In this section we show that, in some cases, our machinery can be used to get locality-preserving reductions between different primitives. That is, we can transform a primitive $\mathscr{F}$ (say one-to-one one-way function) into a different primitive $\mathscr{G}$ (say pseudorandom generator) while preserving the input and output locality of $\mathscr{F}$. Given such a reduction and an implementation of $\mathscr{F}$ with input locality

$\text{in}(n)$ and output locality $\text{out}(n)$ we get an implementation of $\mathscr{G}$ with input locality $\text{in}(n) + O(1)$ and output locality $\text{out}(n) + O(1)$. In particular, if $\mathscr{F}$ can be implemented with constant input locality and constant output locality then so can $\mathscr{G}$.

The general idea is to encode the known construction from $\mathscr{F}$ to $\mathscr{G}$ into a corresponding $\mathbf{Local}_{O(1)}^{O(1)}$ construction. Consider, for example, the Blum-Micali-Yao construction [36, 144] of PRG collection G from one-way permutation f which is defined by $G_z(x) = (f(x), \langle x, z \rangle)$, where $x, z \in \{0, 1\}^n$ (recall that the collection key z is public). Then, for any fixed collection key z the term $\langle x, z \rangle$ is just a fixed function $L_z(x)$ which is linear in x. Hence, by Remark 8.1 it can be encoded by a function $\hat{L}_z(x, r) \in \mathbf{Local}_3^3$. Therefore, the function $\hat{G}_z(x, r) = (f(x), \hat{L}_z(x, r))$ is a (perfect) encoding of G_z and so it forms a reduction from a collection of PRGs to a one-way permutation. This reduction preserves the locality of f. In particular, when $f \in \mathbf{Local}_{\text{in}}^{\text{out}}$ we get a collection of PRGs in $\mathbf{Local}_{\text{in}+1}^{\text{out}}$, assuming that $\text{in} \geq 2$, $\text{out} \geq 3$.[4]

More generally, let $\mathscr{G}$ be a cryptographic primitive whose security is respected by perfect encoding. Suppose that $G(x) = g(x, f(x^{(1)})), \ldots, f(x^{(k)})$ defines a black-box construction of $\mathscr{G}$ from an instance f of a primitive $\mathscr{F}$, where g can be encoded in $\mathbf{Local}_{O(1)}^{O(1)}$ and the concatenation of $x^{(1)}, \ldots, x^{(k)}$ forms a prefix of the input x, i.e., $x = (x^{(1)}, \ldots, x^{(k)}, x^{(k+1)})$. (The function g is fixed by the reduction and does not depend on f.) Then, letting $\hat{g}((x, y_1, \ldots, y_k), r)$ be a perfect $\mathbf{Local}_{O(1)}^{O(1)}$ encoding of g, the function $\hat{G}(x, r) = \hat{g}((x, f((x^{(1)})), \ldots, f(x^{(k)})), r)$ perfectly encodes G, and, hence, defines a black-box locality-preserving reduction from a $\mathscr{G}$ to $\mathscr{F}$.

It turns out that several known cryptographic reductions are of the above form where the function g is a random public linear function (e.g., g is used to extract hardcore bits using Goldreich-Levin [77], or as a pairwise independent hash function). Since linear functions can be encoded by $\mathbf{Local}_3^3$ functions (Remark 8.1) we get a locality-preserving reduction for every fixed choice of the linear function. This results in a locality-preserving transformation from $\mathscr{F}$ to a collection of $\mathscr{G}$. In the following lemma we instantiate this approach with several cryptographic constructions.

Lemma 8.7 *Let* $\text{in}(n) \geq 2$, $\text{out}(n) \geq 3$ *be locality parameters. Let* $f \in \mathbf{Local}_{\text{in}(n)}^{\text{out}(n)}$. *Then,*

1. f *is distributionally one-way* $\Rightarrow \exists$ *collection of OWFs in* $\mathbf{Local}_{\text{in}(n)+1}^{\text{out}(n)}$.
2. f *is a regular OWF* $\Rightarrow \exists$ *collection of PRGs in* $\mathbf{Local}_{\text{in}(n)+1}^{\text{out}(n)+1}$.
3. f *is a one-to-one OWF* $\Rightarrow \exists$ *non-interactive commitment scheme such that the sender's computation is in* $\mathbf{Local}_{\text{in}(n)+1}^{\text{out}(n)}$.
4. f *is a one-to-one trapdoor function* $\Rightarrow \exists$ *public-key encryption scheme* (G, E, D) *such that the encryption algorithm* E *is in* $\mathbf{Local}_{\text{in}(n)+1}^{\text{out}(n)}$.
5. f *is a PRG* $\Rightarrow \exists$ *one-time symmetric-key encryption* (E, D) *(with a short key) such that the encryption algorithm* E *is in* $\mathbf{Local}_{\text{in}(n)+1}^{\text{out}(n)+1}$.

---

[4] The input locality is $\text{in} + 1$ since the input locality of the x's in $\hat{L}_z$ is only 1. See Remark 8.1.

We note that the techniques of [18] can be used to strengthen the first two items and derive a single OWF (resp., PRG). Similarly, these techniques can be used to derive a non-interactive commitment with no set-up algorithm.

Proof (1) Let f be a distributional OWF. We will employ the reduction of [90] which can be written as $F_z(x) = (f(x), L_z(x))$ where L_z is a linear function (originally, L_z is a projection of a pairwise independent hash function). In [90] it was shown that the collection F is weakly one-way (see also p. 96 of [70]). By Remark 8.1, L_z can be encoded by a function $\hat{L}_z(x, r) \in \textbf{Local}_3^3$. Hence, the encoding $\hat{F}_z(x, r) = (f(x), \hat{L}_z(x, r))$ forms a collection of weak OWF. Since the input locality of the original variables (the x's) in $\hat{L}_z$ is only 1, the input locality of $\hat{F}_z$ is only $\text{in}(n) + 1$. Finally, we can transform $\hat{F}$ to a strong (i.e., standard) OWF by applying $\hat{F}$ to polynomially many independent inputs (cf. [144], Theorem 2.3.2 of [70]). This step does not increase the input locality nor the output locality of the reduction. Hence, we get a collection of OWF in $\textbf{Local}_{\text{in}(n)+1}^{\text{out}(n)}$.

(2) We rely on the PRG construction from [86]. When this construction is applied to a regular OWF f, it involves only the computation of universal hash functions and hardcore bits, and therefore can be written as $G_z(x) = L_z(x, f(x^{(1)})), \ldots, f(x^{(k)}))$, where L_z is a linear function (see Theorem 5.1.4 of [86]).[5] Hence, by Remark 8.1, we can encode the reduction by a function $\hat{L}_z(x, y_1, \ldots, y_k, r)$ in $\textbf{Local}_3^3$. If f is in $\textbf{Local}_{\text{in}(n)}^{\text{out}(n)}$ then the input locality of the x's in G_z is $\text{in}(n) + 1$, and the output locality of the outputs that involve x is $\text{out}(n) + 1$. (The random inputs r affect 3 outputs and participate in outputs that depend on 3 inputs.)

(3,4) We rely on the construction of [34] (instantiated with the Goldreich-Levin hardcore predicate [77]): to commit to b using the randomness s, z we compute $\text{COM}(b; s, z) = (f(s), \langle s, z \rangle + b, z)$. Since the degree and rank of the function $g(b, s, z) = (\langle s, z \rangle + b, z)$ are both 2, we can apply the same argument used in (1). The same construction results in a public-key encryption scheme when f is a one-to-one trapdoor function. (See Proposition 5.3.14 of [72].) Note that in both cases, encrypting (resp. committing to) a long message can be done by applying the basic construction to every bit separately (with independent fresh randomness). This extension preserves the locality of the 1-bit schemes.

(5) In Construction 5.1 it was shown how to transform a PRG with minimal (1-bit) stretch into a one-time semantically secure private-key encryption that al-

---

[5]In more detail, suppose that we have a OWF $f : \{0, 1\}^n \to \{0, 1\}^m$ which is t-to-one, where $t = t(n)$ is computable in polynomial time. Then, the construction of [86] (see also [87]) can be described in two steps: First we define a collection $g_{y,r}(x) = (f(x), h_y(x), \langle x, r \rangle)$ where $\{h_y\}$ is a collection of pairwise independent hash functions that map n bits to $\log(t) + 2$ bits. This function is shown to have large "pseudoentropy", that is, when y, r and x are randomly chosen, the bit $\langle x, r \rangle$ has low entropy given the values of $g(x), h_y(x), y$ and r, but it is computationally unpredictable. Then, in the second step we use many instances of g to construct a PRG. That is, we define the function $G_{w, \vec{y}, \vec{r}}(\vec{x}) = h'_w(g_{y^{(1)}, r^{(1)}}(x^{(1)}), \ldots, g_{y^{(k)}, r^{(k)}}(x^{(k)}))$, where $\{h'_w\}$ is a collection of pairwise independent hash functions of output length ℓ, and k, ℓ are some explicit functions of n, m and t. Hence, when $z = (w, \vec{y}, \vec{r})$ is fixed, and the hash functions are implemented via affine transformations, $G_z(\vec{x})$ can be written as $L_z(\vec{x}, f(x^{(1)}), \ldots, f(x^k))$ where L_z is an affine function.

lows us to encrypt messages whose length is polynomially longer than the key length (for any arbitrary polynomial). Specifically, the encryption algorithm was defined as follows: $E_k(x, (s_1, \ldots, s_{\ell-1})) \overset{\text{def}}{=} (G(k) \oplus s_1, G(s_1) \oplus s_2, \ldots, G(s_{\ell-2}) \oplus s_{l-1}, G(s_{\ell-1}) \oplus x)$, where $k \leftarrow U_n$ is the private key, x is a $(k + \ell)$-bit plaintext and $s_i \leftarrow U_{n+i}$ serve as the coin tosses of E. When the PRG is in $\mathbf{Local}_{\text{in}(n)}^{\text{out}(n)}$ we get an encryption in $\mathbf{Local}_{\text{in}(n)+1}^{\text{out}(n)+1}$. $\qquad\square$

We note that the theorem holds even when f is a collection.

8.5 Negative Results for Cryptographic Primitives

In this section we show that cryptographic tasks which require some form of "non-malleability" cannot be performed by functions with low input locality. This includes MACs, signatures and non-malleable encryption schemes (e.g., CCA2 secure encryptions). We prove our results in the private-key setting (i.e., for MAC and symmetric encryption). This makes them stronger as any construction that gains security in the public-key setting is also secure in the private-key setting. We will also prove that there is no PRG in $\mathbf{Local}_2$ and therefore the results of Sect. 8.4.2 are optimal.

8.5.1 Basic Observations

Let $f : \{0, 1\}^n \rightarrow \{0, 1\}^{s(n)}$ be a function and let $s = s(n)$. For $i \in [n]$ and $x \in \{0, 1\}^n$, we let $Q_i(x) \subseteq [s]$ be the set of indices in which $f(x)$ and $f(x_{\oplus i})$ differ. (Recall that $x_{\oplus i}$ denotes the string x with the i-th bit flipped.) We let $Q_i^n \overset{\text{def}}{=} \bigcup_{x \in \{0,1\}^n} Q_i(x)$, equivalently, Q_i^n is the set of output bits which are affected by the i-th input bit. From now on, we omit the superscript n whenever the input length is clear from the context. We show that, given oracle access to f, we can efficiently approximate the set Q_i for every i.

Lemma 8.8 *There exists a probabilistic algorithm A that, given oracle access to $f : \{0, 1\}^n \rightarrow \{0, 1\}^s$, an index $i \in [n]$ and an accuracy parameter ε, outputs a set $Q \subseteq Q_i$ such that $\Pr_x[Q_i(x) \not\subseteq Q] \leq \varepsilon$, where the probability is taken over the coin tosses of A and the choice of x (which is independent of A). Moreover, when $\{f : \{0, 1\}^n \rightarrow \{0, 1\}^{s(n)}\}$ is an infinite collection of functions the time complexity of A is polynomial in n, $s(n)$ and $1/\varepsilon(n)$. In particular, if $s(n) = \text{poly}(n)$, then for every constant c, one can reduce the error to n^{-c} in time $\text{poly}(n)$.*

Proof Let $t = \ln(2s/\varepsilon)$ and $\alpha = \varepsilon/(2s)$. The algorithm A constructs the set Q iteratively, starting from an empty set. In each iteration, A chooses uniformly and independently at random a string $x \in \{0, 1\}^n$ and adds to Q the indices for which $f(x)$

and $f(x_{\oplus i})$ differ. After t/α iterations A halts and outputs Q. Clearly, $Q \subseteq Q_i$. Let $p_j \stackrel{\text{def}}{=} \Pr_x[j \in Q_i(x)]$. We say that j is common if $p_j > \alpha$. Then, if j is common we have

$$\Pr[j \notin Q] \le (1 - p_j)^{t/\alpha} \le (1 - \alpha)^{t/\alpha} \le \exp(-t) = \varepsilon/(2s).$$

Since $|Q_i| \le s$, there are at most s common j's and thus, by a union bound, the probability that A misses a common j is at most $\varepsilon/2$. On the other hand, for a random x, the probability that $Q_i(x)$ contains an uncommon index is at most $s \cdot \alpha = \varepsilon/2$. Hence, we have $\Pr_x[Q_i(x) \not\subseteq Q] \le \varepsilon$, which completes the proof. □

Our negative results are based on the following simple observation.

Lemma 8.9 *Let $f : \{0, 1\}^n \rightarrow \{0, 1\}^{s(n)}$ be a function in $\mathbf{Local}_{\text{in}(n)}$. Then, there exists a probabilistic polynomial-time algorithm that given an input $(y = f(x), i, Q_i^n, 1^n)$, where $x \in \{0, 1\}^n$ and $i \in [n]$, outputs a string $y' = f(x_{\oplus i})$ with probability at least $2^{-\text{in}(n)}$. In particular, when $\text{in}(n) = O(\log(n))$, the success probability is $1/\text{poly}(n)$.*

Proof Fix n and let $s = s(n)$ and $Q_i = Q_i^n$. By definition, y and y' may differ only in the indices of Q_i. Hence, we may randomly choose y' from a set of size $2^{|Q_i|} \le 2^{\text{in}(n)}$, and the lemma follows. □

Note that the above lemma generalizes to the case in which, instead of getting the set Q_i^n, the algorithm A gets a set Q_i' that satisfies $Q_i(x) \subseteq Q_i' \subseteq Q_i^n$.

By combining the above lemmas we get the following corollary.

Corollary 8.2 *Let $f : \{0, 1\}^n \rightarrow \{0, 1\}^{s(n)}$ be a function in $\mathbf{Local}_{\text{in}(n)}$, where $s(n) = \text{poly}(n)$. Then, there exists a probabilistic polynomial-time algorithm A that, given oracle access to f, converts, with probability $(1 - 1/n) \cdot 2^{-\text{in}(n)}$, an image $y = f(x)$ of a randomly chosen string $x \leftarrow U_n$ into an image $y' = f(x_{\oplus 1})$. Namely,*

$$\Pr_x\left[A^f\left(f(x), 1^n\right) = f(x_{\oplus 1})\right] \ge (1 - 1/n) \cdot 2^{-\text{in}(n)},$$

where the probability is taken over the choice of x and the coin tosses of A. In particular, when $l(n) = O(\log(n))$ the algorithm A succeeds with probability $1/\text{poly}(n)$.

Proof First, we use algorithm A_1 of Lemma 8.8 to learn, with accuracy $\varepsilon = 1/n$, an approximation Q_1' of the set Q_1^n. Then, we invoke the algorithm A_2 of Lemma 8.9 on $(f(x), 1, Q_1', 1^n)$ where $f(x)$ is the challenge given to us, and output the result. Let E_1 be the event where $Q_1(x) \subseteq Q_1' \subseteq Q_1$. Since x is uniformly chosen, Lemma 8.8 implies that this event happens with probability larger than $1 - 1/n$. Let E_2 denote the event that A_2 succeeds and outputs $f(x_{\oplus 1})$. Conditioning on the event E_1, the probability of E_2 is at least $2^{-\text{in}(n)}$ (by Lemma 8.9). It follows that our overall success probability is at least $(1 - 1/n) \cdot 2^{-\text{in}(n)}$, and the corollary follows. □

Clearly, one can choose to flip any input bit and not just the first one. Also, we can increase the success probability to $(1 - n^{-c}) \cdot 2^{-\text{in}(n)}$ for any constant c. In the next sections, we will employ Corollary 8.2 to obtain several impossibility results.

8.5.2 MACs and Signatures

Let (G, S, V) be a MAC scheme, where G is a key generation algorithm, the randomized signing function $S(k, \alpha, r)$ computes a signature β on the document α using the key k and randomness r, and the verification algorithm $V(k, \alpha, \beta)$ verifies that β is a valid signature on α using the key k. The scheme is secure (unforgeable) if it is infeasible to forge a signature in a chosen message attack. Namely, any efficient adversary that gets oracle access to the signing process $S(s, \cdot)$ fails to produce a valid signature β on a document α (with respect to the corresponding key k) for which it has not requested a signature from the oracle.[6] The scheme is one-time secure if the adversary is allowed to query the signing oracle only once. One-time secure MACs are known to exist even in an information-theoretic setting. Such schemes do not require any assumption and are secure even against computationally unlimited adversaries.

Suppose that the signature function $S(k, \alpha, r)$ has logarithmic input locality (i.e., $S(k, \alpha, r) \in \textbf{Local}_{O(\log(|k|))}$). Then, we can use Corollary 8.2 to break the scheme with a single oracle call. First, ask the signing oracle $S(k, \cdot)$ to sign on a randomly chosen document α. Then, use the algorithm of Corollary 8.2 to transform, with probability $1/\text{poly}(n)$, the valid pair (α, β) we received from the signing oracle into a valid pair $(\alpha_{\oplus 1}, \beta')$. (Note that when applying Corollary 8.2 we let $S(\cdot, \cdot, \cdot)$ play the role of f.)

Now, suppose that for each *fixed* key $k \in \{0, 1\}^n$ the signature function $S_k(\alpha, r) = S(k, \alpha, r)$ has input locality $\text{in}(n)$. In this case we cannot use Corollary 8.2 directly. The problem is that we cannot apply Lemma 8.8 to learn the set Q_i (i.e., the set of output bits which are affected by the i-th input bit of $f = S_k(\cdot, \cdot)$) since we do not have a full oracle access to S_k. (In particular, we do not see or control the randomness used in each invocation of S_k.) However, we can guess the set Q_i and then apply Lemma 8.9. This attack succeeds with probability $(1/\binom{s(n)}{\text{in}(n)}) \cdot 2^{-\text{in}(n)}$ where $s(n)$ is the length of the signature (and so is polynomial in n). When $\text{in}(n) = c$ is constant, the success probability is $1/\Theta(s(n)^c) = 1/\text{poly}(n)$ and therefore, in this case, we break the scheme.[7] We summarize.

Theorem 8.7 *Let (G, S, V) be a MAC scheme. If $S(k, \alpha, r) \in \textbf{Local}_{O(\log(|k|))}$ or $S_k(\alpha, r) \in \textbf{Local}_{O(1)}$ for every k, then the scheme is not one-time secure.*

---

[6]When querying the signing oracle, the adversary chooses only the message and is not allowed to choose the randomness which the oracle uses to produce the signature.

[7]When the locality $\text{in}(n)$ of S_k is logarithmic (for every fixed key k), this approach yields an attack that succeeds with probability $1/n^{\Theta(\log(n))}$.

Remarks on Theorem 8.7

1. Theorem 8.7 is true even if *some* bit of α has low input locality. This observation also holds in the case of a non-malleable encryption scheme.
2. If we have access to the verification oracle (for example, in the public-key setting where (G, S, V) is a digital-signature scheme), we can even break the scheme in a stronger sense. Specifically, we can forge a signature to *any* target document given a single signature to, say, 0^n. To see this note that, given a signature β of the document α, we can *deterministically* find a signature β' of the document $\alpha_{\oplus i}$ by checking all the polynomially many candidates. Hence, we can apply this procedure (at most) n times and gradually transform a given signature of some arbitrary document into a signature of any target document. Therefore, such a scheme is universally forge-able.
3. A *weaker* version of Theorem 8.7 still holds even when the input locality of the signing algorithm is *logarithmic* with respect to any *fixed* key (i.e., when $S_k(\alpha, r) \in \mathbf{Local}_{O(\log(|k|))}$ for every k). In particular, we can break such MAC schemes assuming that we are allowed to ask for several signatures that were produced with some fixed (possibly unknown) randomness. In such a case, we use Lemma 8.8 to (approximately) learn the output bits affected by, say, the first input bit, and then apply Lemma 8.9 to break the scheme. This attack rules out the existence of a *deterministic* MAC scheme for which $S_k(\alpha) \in \mathbf{Local}_{O(\log(|k|))}$ for every k.

Theorem 8.7 is tight since if the signing algorithm is allowed to use super-constant input locality (for every fixed key), then there exists a one-time secure MAC. Formally, we have.

Lemma 8.10 *Let* $\mathrm{in}(n)$ *be a locality function and* $s(n)$ *be a signature length function. Then, there exists a MAC scheme* (G, S, V) *which cannot be broken by any (computationally unlimited) adversary via a one-time attack with probability larger than* $1/\binom{s(n)}{\mathrm{in}(n)}$. *Moreover,* $S_k(\alpha, r) \in \mathbf{Local}_{\mathrm{in}(n)}$ *for every fixed key* k. *In particular, by setting* $s(n) = \Omega(n)$, *we get superpolynomial security for any super-constant locality.*

Proof Given the length of the signature $s(n)$ and the locality parameter $\mathrm{in}(n)$, we construct the following scheme:

- **Key Generation**. Choose a random $s(n) \times n$ binary matrix M by selecting each of the n columns uniformly and independently from the set of all $s(n)$-bit vectors whose Hamming weight is exactly $\mathrm{in}(n)$. In addition, uniformly choose an $s(n)$ bit vector v.
- **Signature**. To sign compute $S_{M,v}(\alpha) = M \cdot \alpha + v$.
- **Verification**. To verify that (α, β) is valid check if $M \cdot \alpha + v = \beta$.

First, note that each input variable affects at most $\mathrm{in}(n)$ output bits as any column of M has at most $\mathrm{in}(n)$ ones. We move on to prove that the scheme is secure. Let $s = s(n)$. We begin by showing that if the adversary does not query the signing algorithm at all, then she cannot forge a signature. Fix $\alpha \in \{0, 1\}^n$ and $\beta \in \{0, 1\}^s$. Then we can write

$$\Pr_{M,v}[M\alpha + v = \beta] = \Pr_{M,v}[v = \beta - M\alpha] = 2^{-s},$$

where the last equality holds since for every fixed M we have $\Pr_v[v = \beta - M\alpha] = 2^{-s}$.

Suppose that the adversary used his single oracle query to learn the signature $\beta^{(1)}$ to some document $\alpha^{(1)}$. We show that the probability that the adversary finds a signature on any other document $\alpha^{(2)} \neq \alpha^{(1)}$ is at most $1/\binom{s(n)}{\ln(n)}$. Indeed, fix some $\alpha^{(1)} \neq \alpha^{(2)} \in \{0, 1\}^n$ and $\beta^{(1)}, \beta^{(2)} \in \{0, 1\}^s$. We will prove that

$$\Pr_{M,v}\left[M\alpha^{(2)} + v = \beta^{(2)} | M\alpha^{(1)} + v = \beta^{(1)}\right] \leq \frac{1}{\binom{s(n)}{\ln(n)}}.$$

First note that $M\alpha^{(1)} + v = \beta^{(1)}$ if and only if $v = \beta^{(1)} - M\alpha^{(1)}$. Hence, by letting $\alpha = \alpha^{(2)} - \alpha^{(1)}$ and $\beta = \beta^{(2)} - \beta^{(1)}$, we have

$$\Pr_{M,v}\left[M\alpha^{(2)} + v = \beta^{(2)} | M\alpha^{(1)} + v = \beta^{(1)}\right] = \Pr_M[M\alpha = \beta].$$

Observe that $\alpha \neq 0^n$ (since $\alpha^{(1)} \neq \alpha^{(2)}$). Assume, without loss of generality, that α_1, the first bit of α is 1 (otherwise, permute α and the columns of M). Let M_i denote the i-th column of M. Then,

$$\Pr_M[M\alpha = \beta] = \Pr_M\left[M_1 = \beta - \sum_{i=2}^n \alpha_i \cdot M_i\right].$$

We complete the proof by noting that M_1 is distributed uniformly over a set of size $\binom{s(n)}{\ln(n)}$ independently of the other columns, and thus the above term is bounded by $1/\binom{s(n)}{\ln(n)}$.   □

8.5.3 Non-malleable Encryption

Let (G, E, D) be a private-key encryption scheme, where G is a key generation algorithm, the encryption function $E(k, m, r)$ computes a ciphertext c encrypting the message m using the key k and randomness r, and the decryption algorithm $D(k, c, r)$ decrypts the ciphertext c that was encrypted under the key k. Roughly speaking, non-malleability of an encryption scheme guarantees that it is infeasible to modify a ciphertext c into a ciphertext c' of a message related to the decryption of c.

Theorem 8.8 *Let (G, E, D) be a private-key encryption scheme. If $E(k, m, r) \in$* **Local**$_{O(\log(|k|))}$ *or $E_k(m, r) \in$* **Local**$_{O(1)}$ *for every k, then the scheme is malleable* *with respect to an adversary that has no access to either the encryption oracle or the decryption oracle. If (G, E, D) is a public-key encryption scheme and $E_k(m, r) \in$* **Local**$_{O(\log(|k|))}$ *for every k, then the scheme is malleable.*

Proof The proof is similar to the proof of Theorem 8.7. Let n be the length of the key k, $p = p(n), m = m(n)$, and $s = s(n)$ be the lengths of the message x, randomness r, and ciphertext length c respectively; i.e., $E : \{0, 1\}^n \times \{0, 1\}^p \times \{0, 1\}^m \to \{0, 1\}^s$. Our attacks will use the uniform message distribution $\mathcal{M} = U_p$ and the relation R for which $(x, x') \in R$ if and only if x and x' differ only in their first bit.

Suppose that the encryption function $E(k, x, r)$ has logarithmic input locality (i.e., $E(k, x, r) \in$ **Local**$_{O(\log(|k|))}$). Then, by Corollary 8.2, we can break the scheme by transforming, with noticeable probability, the challenge ciphertext c into a ciphertext c' such that the corresponding plaintexts differ only in their first bit. Clearly, the probability for this relation to hold with respect to $\tilde{c}$ which is a ciphertext of a random plaintext is negligible. Hence, we break the scheme.

Now, suppose that for each fixed key $k \in \{0, 1\}^n$ the encryption function $E_k(x, r) = E(k, x, r)$ has input locality $in(n)$. In this case we guess the set Q_1 and then apply Lemma 8.9. This attack succeeds with probability $(1/\binom{s(n)}{in(n)}) \cdot 2^{-in(n)}$. When $in(n)$ is constant, the success probability is $1/\text{poly}(n)$ and therefore, in this case, the scheme is broken.

We move on to the case in which the input locality of E_k is logarithmic. The previous attack succeeds in this case with probability $1/n^{\Theta(\log(n))}$. However, we can improve this to $1/\text{poly}(n)$ if we have *stronger* access to the encryption oracle. In particular, we should be able to get several ciphertexts that were produced with some fixed (possibly unknown) randomness. In such a case, we use Lemma 8.8 to (approximately) learn the output bits affected by, say, the first input bit, and then apply Lemma 8.9 to break the scheme. The public-key setting is a special case in which this attack is feasible as we get a full access to the randomness of the encryption oracle. $\square$

8.5.4 The Impossibility of Implementing a PRG in Local$_2$

We prove that there is no PRG in **Local**$_2$ and thus the PRG constructed in Sect. 8.4.2 has optimal input locality as well as optimal output locality.

Lemma 8.11 *Let $G : \{0, 1\}^n \to \{0, 1\}^{s(n)}$ be a polynomial-time computable function in* **Local**$_2$ *where $s(n) > n$. Then, there exists polynomial-size circuit family $\{A_n\}$ that given $z \in \text{Im}(G)$ reads some subset $S \subset [s(n)]$ of z's bits and outputs (S, k, z_k) for some $k \notin S$.*

Proof Fix n and let $s = s(n)$. Define $H_G = ((\text{Out} = [s], \text{In} = [n]), E)$ to be the bipartite graph whose edges correspond to the input-output dependencies in G; that is, (i, j) is an edge if and only if the i-th output bit of G depends on the j-th input bit. Since G is in $\textbf{Local}_2$ the average degree of output vertices is $2n/s$ which is smaller than 2 (as $s > n$). The circuit A_n implements the following procedure:

1. Initialize S and T to be empty sets, let $H \leftarrow H_G$ and let $x = 0^n$.
2. If there exists an output $k \in \text{Out}$ which is not connected to any input in the graph H, then halt and predict z_k by computing $G(x)_k$.
3. Otherwise, there exists an output $j \in \text{Out}$ which depends on a single input bit $i \in \text{In}$ in the graph H (since the average out degree is smaller than 2). Add j to S and add i to T. Let $x_i = 0$ if $G(x) = z_j$, and 1 otherwise. Remove i and j from the graph H.
4. Goto 2.

The procedure stops after at most n steps since there are only $n < s$ inputs. The correctness follows by noting that: (1) in each iteration $x_T = x_T^\star$ where $x^\star$ is the preimage of z under G; and (2) the k-th output bit depends only on the input bits which are indexed by T. To see (1) observe that in each iteration all the output bits which are indexed by S depend only on the input bits which are indexed by T. $\square$

We can now conclude that there is no PRG in $\textbf{Local}_2$.

Corollary 8.3 *There is no PRG in* $\textbf{Local}_2$.

Proof Assume, towards a contradiction, that $G : \{0,1\}^n \rightarrow \{0,1\}^{s(n)}$ is a PRG in $\textbf{Local}_2$. Fix n and let $s = s(n)$. Let A_n be the adversary defined in Lemma 8.11. Then, we define an adversary B_n that given $z \in \{0,1\}^s$ invokes A_n and checks whether A_n predicts z_k correctly. By Lemma 8.11, when $z \in \text{Im}(G)$ the adversary B_n always outputs 1. However, when z is a random string the probability that B_n outputs 1 is at most $1/2$. Indeed, if $B_n(z) = 1$ for some $z \in \{0,1\}^s$ then $B_n(z_{\oplus k}) = 0$, where k is the bit that A_n predicts when reading z (and $z_{\oplus i}$ denote the string z with the i-th bit flipped). Hence, A_n errs on at least half of the strings in $\{0,1\}^s$, and so it distinguishes $G(U_n)$ from U_s with advantage $1/2$. $\square$

The above corollary can be extended to rule out the existence of a *collection* of PRGs with input locality 2. Note that when G is chosen from a collection, the graph H_G might not be available to the adversary constructed in Lemma 8.11. However, a closer look at this lemma shows that, in fact, the adversary does not need an explicit description of H_G; rather, it suffices to find an approximation of H_G (in the sense of Lemma 8.8). As shown in Lemma 8.8, such an approximation can be found efficiently (with, say, $1/n$ error probability) given an (oracle) access to G. This modification also shows that such a PRG can be broken by a *uniform* adversary.

The above negative result does not rule out the existence of a OWF in $\textbf{Local}_2$, which is left as an open problem.

8.6 Negative Results for Randomized Encodings

In the following, we prove some negative results regarding randomized encoding with low input locality.

In Sect. 8.6.1, we provide a necessary condition for a function to have such an encoding. We use this condition to prove that some simple ($\mathbf{NC}^0$) functions cannot be encoded by functions having sublinear input locality (regardless of the complexity of the encoding). This is contrasted with the case of constant output locality, where it is shown (Chap. 4) that *every* function f can be encoded by a function $\hat{f}$ with output locality 4 (and complexity polynomial in the size of a branching program that computes f).

In Sect. 8.6.2 we show that, although linear functions do admit efficient constant-input encoding, they do not admit an efficient *universal* constant-input encoding. That is, one should use different decoders and simulators for different linear functions. This is contrasted with previous constructions of randomized encoding with constant output locality (Chap. 4) which give a (non-efficient) universal encoding for the class of all functions $f : \{0, 1\}^n \rightarrow \{0, 1\}^l$ as well as an efficient universal encoding for classes such as all linear functions or all size-s BPs (where s is polynomial in n).

These results hold in the case of perfect encoding as well as in the more liberal setting of statistically correct and statistically (or even computationally) private encodings in which the simulator and decoder are allowed to err.

8.6.1 A Necessary Condition for Encoding with Low Input Locality

Let $f : \{0, 1\}^n \rightarrow \{0, 1\}^l$ be a function. Define an undirected graph $G_i(f)$ over $\mathrm{Im}(f)$ such that there is an edge between the strings y and y' if there exists $x \in \{0, 1\}^n$ such that $f(x) = y$ and $f(x_{\oplus i}) = y'$. Note that two vertices which lie in the same connected component of $G_i(f)$ differ only in the indices which are affected by the i-th input. Hence, when f has low input locality in, the size of each component of $G_i(f)$ is at most $2^{|\mathrm{in}|}$. It turns out that a similar restriction also holds when f is encoded by a function $\hat{f}$ with low input locality, even when f itself has large input locality. Specifically, in Sects. 8.6.1.1 and 8.6.1.2 we will prove the following theorem:

Theorem 8.9 *Let* $f : \{0, 1\}^n \rightarrow \{0, 1\}^l$ *be a function which is encoded by a function* $\hat{f} : \{0, 1\}^n \times \{0, 1\}^m \rightarrow \{0, 1\}^s$ *in* $\mathbf{Local}_{\mathrm{in}}$. *Then,*

1. *If* $\hat{f}$ *is a perfect encoding then for every* $1 \leq i \leq n$ *the size of the connected components of* $G_i(f)$ *is at most* 2^{in}.
2. *If* $\hat{f}$ *is a* δ-*correct and* ε-*private encoding then for every* $1 \leq i \leq n$ *the degree of each vertex of* $G_i(f)$ *is at most* $\frac{n}{-\log(\delta+\varepsilon)} \cdot 2^{\mathrm{in}}$. *In particular, if* $\varepsilon + \delta < 0.9$ *then the degree is bounded by* $7n2^{\mathrm{in}}$.

We will actually show that the second conclusion holds even when the privacy is relaxed to be $\varepsilon(n)$-computational as long as in$(n) \leq O(\log n)$. Also note that the theorem is meaningful as long as the sum of ε and δ is upper bounded away from 1. This limitation is rather weak since one typically requires ε and δ to be negligible in n.

Theorem 8.9 shows that even some very simple functions do not admit an encoding with constant input locality. Consider, for example, the function

$$f(x_1, \ldots, x_n) = x_1 \cdot (x_2, \ldots, x_n) = (x_1 \cdot x_2, x_1 \cdot x_3, \ldots, x_1 \cdot x_n).$$

For every $y \in \text{Im}(f) = \{0, 1\}^{n-1}$ it holds that $f(1, y) = y$ and $f(0, y) = 0^{n-1}$. Hence, every vertex in G_1 is a neighbor of 0^{n-1} and the size of the connected component of G_1 is 2^{n-1}. Thus, the input locality of x_1 in any perfect encoding (respectively, computational encoding) of this function is at least $n - 1$ (respectively, $n - 2 - \log n$ for sufficiently large n). Note that this matches the results of Sect. 8.3 since rank$(x_1) = n - 1$.

As an additional example, consider a random function that maps n-bit strings to $(1 - \varepsilon)n$-bit strings for some small constant $\varepsilon > 0$. We argue that, except with exponentially small probability, all G_i's have maximal degree of size $2^{\Omega(n)}$ and thus cannot be encoded with input locality smaller than $\Omega(n)$. To see this fix $i \in [n]$ and fix the value of f on all strings whose i-th bit is 0. By the pigeonhole principle, at least $2^{\varepsilon n - 1}$ of these strings are mapped to a single image $y \in \{0, 1\}^{(1-\varepsilon)n}$. Call this set of strings X_0. Fix some set X_1 of $2^{\varepsilon' n}$ strings x such that $x_{\oplus i} \in X_0$ where $\varepsilon' < (1 - \varepsilon)/2$. It is not hard to see that, except with exponentially small probability over the choice of $f(X_1)$, the size of $\text{Im}(f(X_1))$ is at least $2^{\Omega(n)}$. (Indeed, we throw $N = 2^{\varepsilon' n}$ balls into $2^{(1-\varepsilon)n} \gg N^2$ bins and so, with all but exponentially small probability, there will be no collisions.) Finally, observe that the node y in G_i is connected to all the strings in $\text{Im}(f(X_1))$ and so it has exponentially many neighbors.

8.6.1.1 Proof of Theorem 8.9 for Perfect Encoding

Fix $i \in [n]$ and let $G = G_i(f)$. Let $\hat{f} : \{0, 1\}^n \times \{0, 1\}^m \to \{0, 1\}^s$ be a perfectly correct and private randomized encoding of $f : \{0, 1\}^n \to \{0, 1\}^l$ with decoder B and simulator S. Let $Q \subseteq \{1, \ldots, s\}$ be the set of output bits in $\hat{f}$ which are affected by the input variable x_i. Namely, $j \in Q$ iff $\exists x \in \{0, 1\}^n, r \in \{0, 1\}^m$ such that the strings $\hat{f}(x, r)$ and $\hat{f}(x_{\oplus i}, r)$ differ on the j-th bit.

We begin with the following claims.

Claim 8.2 *Let $y, y' \in \text{Im}(f)$ be adjacent vertices in G_i. Then, for every $\hat{y} \in$ support$(S(y))$ there exists $\hat{y}' \in$ support$(S(y'))$ which differs from $\hat{y}$ only in indices which are in Q.*

Proof Let $x \in \{0, 1\}^n$ be an input string for which $f(x) = y$ and $f(x_{\oplus i}) = y'$. Fix some $\hat{y} \in$ support$(S(y))$. Then, by perfect privacy, there exists some $r \in \{0, 1\}^m$ for which $\hat{y} = \hat{f}(x, r)$. Let $\hat{y}' = \hat{f}(x_{\oplus i}, r)$. By the definition of Q, the strings $\hat{y}$ and $\hat{y}'$ differ only in indices which are in Q. Also, by the perfect privacy of $\hat{f}$, we have that $\hat{y}' \in \text{support}(S(y'))$ and the claim follows.   □

Claim 8.3 *Let* $y \in \text{Im}(f)$ *and let* $\hat{y} \in \text{Im}(\hat{f})$. *Then,* $y = B(\hat{y})$ *if and only if* $\hat{y} \in \text{support}(S(y))$.

Proof Let $x \in f^{-1}(y)$. By perfect correctness, $y = B(\hat{y})$ iff $\hat{y} \in \text{support}(\hat{f}(x, U_m))$. By perfect privacy, $\text{support}(\hat{f}(x, U_m)) = \text{support}(S(f(x))) = \text{support}(S(y))$, and the claim follows.   □

We can now prove the first part of Theorem 8.9. The idea is to label each vertex y of G by a distinct string $\hat{y} \in \text{Im}(\hat{f})$ and to show that the vertices of each connected component are labeled by a small set of strings. Specifically, we show that if u and v are in the same connected component then their labels differ only in the indices which are in Q. It follows that the size of such a component is bounded by $2^{|Q|}$.

Proof of Theorem 8.9 part 1 Fix $u \in \text{Im}(f)$ and let $\hat{u} \in \{0, 1\}^s$ be some arbitrary element of $\text{support}(S(u))$. Let $Z \stackrel{\text{def}}{=} \{z \in \{0, 1\}^s \mid z_i = \hat{u}_i, \forall i \in [s] \setminus Q\}$. That is, $z \in Z$ if it differs from $\hat{u}$ only in indices which are in Q. To prove the claim, we define an onto mapping from Z (whose cardinality is $2^{|Q|}$) to the members of the connected component of u. The mapping is defined by applying the decoder B of $\hat{f}$, namely $z \to B(z)$. (Assume, w.l.o.g., that if the decoder is invoked on a string z which is not in $\text{Im}(\hat{f})$ then it outputs $\perp$.) Let $v \in \text{Im}(f)$ be a member of the connected component of u. We prove that there exists $z \in Z$ such that $v = B(z)$.

The proof is by induction on the distance (in edges) of v from u in the graph G. In the base case when the distance is 0, we let $z = \hat{u}$ and, by perfect correctness, get that $B(\hat{u}) = u$. For the induction step, suppose that the distance is $i > 1$. Then, let w be the last vertex in a shortest path from u to v. By the induction hypothesis, there exists a string $\hat{w} \in Z$ for which $w = B(\hat{w})$. Hence, by Claim 8.3, $\hat{w} \in \text{support}(S(w))$. Since v and w are neighbors, we can apply Claim 8.2 and conclude that there exists $\hat{v} \in \text{support}(S(v))$ which differs from $\hat{w}$ only in indices which are in Q. Since $\hat{w} \in Z$ it follows that $\hat{v}$ is also in Z. Finally, by Claim 8.3, we have that $B(\hat{v}) = v$ which completes the proof.   □

8.6.1.2 Proof of Theorem 8.9 for Statistical and Computational Encoding

In the following, we will keep the notation of the previous section but relax $\hat{f}$: $\{0, 1\}^n \times \{0, 1\}^m \to \{0, 1\}^s$ to be a δ-correct and ε-private randomized encoding of f.

We say that a string $\hat{y} \in \{0, 1\}^s$ is *good* for x if there exists a string $\hat{u}$ such that: (1) $\hat{u}$ differs from $\hat{y}$ only in indices which are in Q; and (2) $B(\hat{u}) = f(x_{\oplus i})$, where B is the decoder of $\hat{f}$.

Claim 8.4 *For every $x \in \{0,1\}^n$ a string $\hat{y}$ which is chosen from the distribution $S(f(x))$ will be good with probability $1 - \delta - \varepsilon$, where S is a simulator for the encoding.*

Proof Fix x. Consider the imaginary experiment where $\hat{y}$ is chosen from the distribution $\hat{f}(x, r)$ where $r \leftarrow U_m$. Let $\hat{y}' = \hat{f}(x_{\oplus i}, r)$. Clearly, $\hat{u}$ differs from $\hat{y}$ only in indices which are in Q. Furthermore, by the correctness of the encoding $\hat{u}$ decodes to $f(x_{\oplus i})$ with probability at least $1 - \delta$ (since r is uniformly distributed). Hence, in our imaginary experiment $\hat{y}$ is good for x with probability $1 - \delta$. Finally, privacy guarantees that $\| \hat{f}(x, U_m) - S(y) \| \leq \varepsilon$ and therefore the probability that $\hat{y}$ is good for x in the real experiment is at least $1 - \delta - \varepsilon$. $\qquad\square$

Lemma 8.12 *For every $y \in \text{Im}(f)$ there exists a set $T_y \subseteq \text{Im}(\hat{f})$ of size at most $\frac{n}{-\log(\delta+\varepsilon)}$ such that for every $x \in f^{-1}(y)$ there exists a good $\hat{y} \in T_y$.*

Proof Fix y and let $X = f^{-1}(y)$. We will construct the set T_y iteratively. We begin with an empty set T_y and with $X_0 = X$. In the i-th iteration we will choose a string $\hat{y}$ which is good for at least $1 - \delta - \varepsilon$ fraction of the entries in X_i and put the remaining x's in X_{i+1}. Since the initial size of X_0 is bounded by 2^n we will need at most $\frac{n}{-\log(\delta+\varepsilon)}$ iterations.

It is left to argue that in each iteration there exists such a good $\hat{y}$. Fix some $X_i \subseteq X$. By Claim 8.4 and the linearity of expectation, a random $\hat{y}$ which is chosen from the distribution $S(y)$ is expected to be good for at least $1 - \delta - \varepsilon$ fraction of the x's of X_i. The existence of such fixed $\hat{y}$ is therefore guaranteed by an averaging argument. $\qquad\square$

We can now prove the second part of Theorem 8.9. The proof is similar to the proof of the first part. However, now we will label each vertex y of G by a small *collection* of strings $\hat{y}_1, \ldots, \hat{y}_k \in \text{Im}(\hat{f})$. We will show that if u and v are neighbors then there exists a corresponding label $\hat{u}_i$ (in the collection of u) and a string $\hat{v}$ such that: (1) $\hat{u}_i$ and $\hat{v}$ differ only in the locations which are indexed by Q; and (2) $\hat{v}$ decodes to v. It follows that the *degree* of G is bounded by $k \cdot 2^{|Q|}$.

Lemma 8.13 *The degree of each vertex of G is at most $\frac{n}{-\log(\delta+\varepsilon)} \cdot 2^{|Q|}$.*

Proof Fix $u \in \text{Im}(f)$ and let $T_u \subseteq \{0,1\}^s$ be a set of size at most $\frac{n}{-\log(\delta+\varepsilon)}$ which satisfies Claim 8.12. Let $Z \stackrel{\text{def}}{=} \{z \in \{0,1\}^s \mid z_{[s]\setminus Q} = \hat{u}_{[s]\setminus Q}, \exists \hat{u} \in T_u\}$. That is, $z \in Z$ if it differs from some $\hat{u} \in T_u$ only in indices which are in Q. To prove the claim, we define an onto mapping from Z to the neighbors of u. The mapping is defined by applying the decoder B of $\hat{f}$, namely $z \to B(z)$.

Let $v \in \text{Im}(f)$ be a neighbor of u. We prove that there exists $z \in Z$ such that $v = B(z)$. Indeed, let x be a preimage of u for which $v = f(x_{\oplus i})$ and let $\hat{u} \in T_u$ be a good string for x. It follows that there exists a string $\hat{v}$ which decodes to v and agrees with $\hat{u}$ on the coordinates $[s] \setminus Q$, which completes the proof. $\qquad\square$

The Computational Setting   The above lemma extends to the case where the encoding is only ε-computational private (and δ-correct) as long as Q is of logarithmic size and the decoder is efficient. To see this note that Claim 8.4 (which is the only place where privacy was used) still holds in the computational setting. Indeed, if the claim does not hold for some infinite family of $\{x_n\}$ then one can efficiently distinguish between the ensembles $S(f(x_n))$ and $\hat{f}(x_n, U_{m(n)})$ with advantage bigger than ε by checking whether a sample $\hat{y}$ is good for x_n. This test is efficiently computable (by a polynomial-size circuit family) as long as Q is sufficiently small and B is efficient.

8.6.2 Impossibility of Universal Encoding for Linear Functions

For a class C of functions that map n-bits into l-bits, we say that C has a universal encoding in the class $\hat{C}$ if there exists a universal simulator S and a universal decoder B such that, for every function $f_z \in C$, there is an encoding $\hat{f}_z \in \hat{C}$ which is private and correct with respect to the simulator S and the decoder B.

We show that, although linear functions do admit encodings with constant input locality, they do not admit such a *universal* encoding. Suppose that the class of linear (equivalently affine) functions had a universal encoding with constant input locality. Then, by the results of Chap. 4, we would have a one-time secure MACs (S, V) whose signing algorithm has constant input locality for every fixed key; i.e., $S_k(\alpha, r) \in \mathbf{Local}_{O(1)}$ for every fixed key k. However, the results of Sect. 8.5.2 rule out the existence of such a scheme. We now give a more direct proof to the impossibility of obtaining a universal encoding with constant input locality for linear functions. The proof is similar to the proofs in Sect. 8.6.1.

Let C be a class of functions that map n bits into l bits. For each input bit $1 \leq i \leq n$, we define a graph G_i over $\bigcup_{f \in C} \mathrm{Im}(f)$ such that there is an edge between the strings y and y' if there exists $x \in \{0, 1\}^n$ and $f \in C$ such that $f(x) = y$ and $f(x_{\oplus i}) = y'$. Namely, $G_i = \bigcup_{f \in C} G_i(f)$, where $G_i(f)$ is the graph defined in Sect. 8.6.1. Suppose that C has universal encoding in $\mathbf{Local}_{in}$ with decoder B and simulator S. That is, for every $f_z \in C$ there exists a perfect randomized encoding $\hat{f} : \{0, 1\}^n \times \{0, 1\}^m \to \{0, 1\}^s$ in $\mathbf{Local}_{in}$ whose correctness and privacy hold with respect to B and S.

Lemma 8.14   *The degree of every vertex in G_i is bounded by* $\binom{s}{in} \cdot 2^{in}$.

Proof   Let y be a vertex of G_i and fix some $\hat{y} \in S(y)$. Let y' be a neighbor of y with respect to some function $f \in C$. Let $Q = Q_i \subseteq \{1, \ldots, s\}$ be the set of output bits in $\hat{f}$ which are affected by the input variable x_i. Then, by the proof of Theorem 8.9 part 1, there exists a set $Z_Q \subseteq \{0, 1\}^s$ of size 2^{in} such that $y' \in \mathrm{Im}(B(Z_Q))$. Hence, we have an onto mapping from $Q \times Z_Q$ to the neighbors of y. Thus, the number of neighbors is at most $\binom{s}{in} \cdot 2^{in}$. $\qquad\qquad\square$

Lemma 8.15 *Let C be the class of linear functions $L : \{0, 1\}^n \to \{0, 1\}^l$ where $l \leq n$. Then, for every $1 \leq i \leq n$ the graph G_i is a complete graph over $\{0, 1\}^l$.*

Proof Consider, for example, the graph $G = G_1$ and fix some $y, y' \in \{0, 1\}^l$. Then, for $\sigma = (0, 1, \ldots, 1)$ and $\sigma_{\oplus 1} = (1, 1, \ldots, 1)$, there exists a linear function $L : \{0, 1\}^n \to \{0, 1\}^l$ for which $y = L(\sigma)$ and $y' = L(\sigma_{\oplus 1})$. To see this, write L as a matrix $M \in \{0, 1\}^{l \times n}$ such that $L(x) = Mx$. Let $M = (M_1, M')$, that is M_1 denotes the leftmost column of M, and M' denotes the matrix M without M_1. Now, we can first solve the linear system $M \cdot \sigma = y$ which is equivalent to $M' \cdot \sigma = y$ and then solve the linear system $M \cdot \sigma_{\oplus 1} = y'$ which is now equivalent to $M_1 = y' - y$. $\square$

Let $l \leq n$. By combining the above claims we conclude that the output complexity s of any universal encoding in **Local**$_{\text{in}}$ for linear functions $L : \{0, 1\}^n \to \{0, 1\}^l$ must satisfy $\binom{s}{\text{in}} \cdot 2^{\text{in}} \geq 2^l$. In particular, when in is constant, the output complexity of the encoding must be exponential in l. A similar bound also holds when the encoding is only 0.45-correct and 0.45-computationally private as long as l is superlogarithmic in n. This can be proven by a straightforward extension of Lemma 8.14 which uses the second part of Theorem 8.9.

8.7 Conclusions and Open Questions

We showed that, under standard intractability assumptions, cryptographic primitives such as OWFs, PRGs and public-key encryption can be computed by functions of constant input locality. On the other hand, primitives that require some form of "non-malleability", such as MACs, signatures and non-malleable encryption schemes, cannot be computed by such functions. An interesting open question is whether collision-resistant hash functions can be realized by functions of constant input locality. It is not hard to see that such functions are extremely vulnerable to near-collision attacks (since if x, x' are "close" in Hamming distance, then so are their images $h(x)$ and $h(x')$). However, it is not clear whether this weakness allows us to find actual collisions. Recently, some positive results, with respect to universal one-way hash functions, were established in [20].

8.8 Addendum: Cryptography with Physical Locality

Computation in the physical world is restricted by the following *spatial locality* constraint: In a single unit of time, information can only travel a bounded distance in space. Hence, a "real-world" notion of parallel-time computation should take into account the distances traveled by signals in an actual embedding of the circuit in physical space. Ultimately, one can hope for a constant parallel-time implementation in which the length of the wires (in some physical embedding of the circuit) is bounded by an absolute constant. Formally, this means that the input-output dependencies graph of the computed function can be embedded on the two-dimensional grid such that the Euclidean distance between each pair of adjacent nodes is bounded by some constant d. This notion of constant spatial locality is strictly stronger than the combination of constant input locality and constant output locality. In particular, the underlying dependencies graph cannot be a good expander, and as a consequence it can be shown that the cryptographic constructions presented in this book (including those in Chap. 8) are not spatially local.

The question of cryptography with constant spatial locality was studied in [18]. On the negative side, it was shown that spatially local functions can be inverted in subexponential time of $2^{O(\sqrt{n})}$, and that PRGs with linear stretch cannot be spatially local. These attacks exploit the fact that the dependencies graph of a spatially local functions is a poor expander which has relatively small separators.

On the positive side, it was shown that, despite the lack of good expansion, some cryptographic computations can still be implemented in this model. Specifically, under coding related intractability assumptions (similar to those presented in Chap. 8), it is possible to implement OWFs, PRGs, commitments, and encryption schemes (both in the public-key and private-key setting) with constant spatial locality. The proof relies on an extension of the encoding from Sect. 8.3. These techniques also provide *explicit* one-way functions and pseudorandom generators with optimal input-locality and output-locality (simultaneously), improving the results of Chaps. 6 and 8 which were limited to *collections*.

Apart from leading to fast parallel implementations, spatially local cryptography can be used to derive strong hardness results regarding simple dynamical systems. Such systems are commonly modeled by a *cellular automaton* (CA): A discrete dynamical system in which cells are placed on a grid and the state of each cell is updated via a local deterministic rule that depends only on the few cells within its close neighborhood. Conway's Game of Life [30] is a famous instance of CA in which each cell is initialized to be either "alive" or "dead", and in each time step each cell interacts with the eight neighboring cells surrounding it and determines whether to live or die based on the number of its living neighbors.

CAs were introduced by von Neumann and Ulam in an attempt to model natural physical and biological systems [118]. Despite their simple structure, CAs were shown to exhibit complex computational and dynamical phenomena such as self-replication, universal computation, synchronization, fractality, and chaos [29, 111, 118, 136, 143]. The ability to generate complex behaviors by using simple basic rules makes CAs a powerful tool: Scientists use them to simulate and analyze real-world systems in nature and society, and engineers view them as massive parallel processing machines that can efficiently solve complicated algorithmic tasks. (See [65] for a survey on CAs and their applications.)

Being a universal computational model, CAs are capable of highly complex behavior. However, it is not clear *how fast* this complexity can evolve and how common it is with respect to all possible initial configurations. In [18] this question is examined from a computational perspective, identifying "complexity" with computational intractability. In particular, [18] uses spatially local OWFs to show that the task of inverting a *single step* of CA is intractable in the *average case*, i.e., when the CA is initialized to a *random configuration*. Similarly, spatially local PRGs are used to show that it is hard to predict the next configuration of a CA based on a partial observation of its current state, even if the current configuration is chosen uniformly at random. Previously, computational intractability was "generated" after a (polynomially) large number of evolution steps, or was restricted to worst-case choices of the initial configuration. Hence, the results of [18] show that evolution of computational intractability can be both *common* and *fast*. (See [18] for details.)

References

1. Agrawal, M., Allender, E., Rudich, S.: Reductions in circuit complexity: an isomorphism theorem and a gap theorem. J. Comput. Syst. Sci. **57**(2), 127–143 (1998)
2. Aiello, W., Hastad, J.: Statistical zero-knowledge languages can be recognized in two rounds. J. Comput. Syst. Sci. **42**, 327–345 (1991)
3. Ajtai, M.: Generating hard instances of lattice problems. In: Proc. 28th STOC, pp. 99–108 (1996). Full version in Electronic Colloquium on Computational Complexity (ECCC)
4. Ajtai, M., Dwork, C.: A public-key cryptosystem with worst-case/average-case equivalence. In: Proc. of 29th STOC, pp. 284–293 (1997)
5. Alekhnovich, M.: More on average case vs approximation complexity. In: Proc. of 44th FOCS, pp. 298–307 (2003)
6. Alekhnovich, M., Hirsch, E.A., Itsykson, D.: Exponential lower bounds for the running time of DPLL algorithms on satisfiable formulas. J. Autom. Reason. **35**(1–3), 51–72 (2005)
7. Alon, N., Babai, L., Itai, A.: A fast and simple randomized parallel algorithm for the maximal independent set problem. J. Algorithms **7**(4), 567–583 (1986)
8. Alon, N., Roichman, Y.: Random Cayley graphs and expanders. Random Struct. Algorithms **5**(2), 271–285 (1994)
9. Applebaum, B.: Randomly encoding functions: a new cryptographic paradigm (invited talk). In: ICITS, pp. 25–31 (2011)
10. Applebaum, B.: Pseudorandom generators with long stretch and low locality from random local one-way functions. In: Proc. of 44th STOC, pp. 805–816 (2012). Full version in ECCC TR11-007
11. Applebaum, B., Barak, B., Wigderson, A.: Public-key cryptography from different assumptions. In: Proc. of 42nd STOC, pp. 171–180 (2010)
12. Applebaum, B., Bogdanov, A., Rosen, A.: A dichotomy for local small-bias generators. In: Proc. of 9th TCC, pp. 1–18 (2012)
13. Applebaum, B., Ishai, Y., Kushilevitz, E.: On One-Way Functions with Optimal Locality. Unpublished manuscript available at http://www.cs.technion.ac.il/~abenny (2005)
14. Applebaum, B., Ishai, Y., Kushilevitz, E.: Computationally private randomizing polynomials and their applications. Comput. Complex. **15**(2), 115–162 (2006). Preliminary version in Proc. of 20th CCC, 2005
15. Applebaum, B., Ishai, Y., Kushilevitz, E.: Cryptography in NC^0. SIAM J. Comput. **36**(4), 845–888 (2006). Preliminary version in Proc. of 45th FOCS, 2004
16. Applebaum, B., Ishai, Y., Kushilevitz, E.: On pseudorandom generators with linear stretch in NC^0. Comput. Complex. **17**(1), 38–69 (2008). Preliminary version in Proc. of 10th RANDOM, 2006
17. Applebaum, B., Ishai, Y., Kushilevitz, E.: Cryptography with constant input locality. J. Cryptol. **22**(4), 429–469 (2009). Preliminary version in Proc. of 27th CRYPTO, 2007

B. Applebaum, *Cryptography in Constant Parallel Time*,
Information Security and Cryptography,
DOI 10.1007/978-3-642-17367-7, © Springer-Verlag Berlin Heidelberg 2014

18. Applebaum, B., Ishai, Y., Kushilevitz, E.: Cryptography by cellular automata or how fast can complexity emerge in nature? In: ICS, pp. 1–19 (2010)
19. Applebaum, B., Ishai, Y., Kushilevitz, E.: How to garble arithmetic circuits. In: Proc. of 52nd FOCS, pp. 120–129 (2011)
20. Applebaum, B., Moses, Y.: Locally computable UOWHF with linear shrinkage. In: Advances in Cryptology: Proc. of EUROCRYPT '13 (2013)
21. Arora, S., Lund, C., Motwani, R., Sudan, M., Szegedy, M.: Proof verification and hardness of approximation problems. J. ACM 45(3), 501–555 (1998). Preliminary version in Proc. of 33rd FOCS, 1992
22. Arora, S., Safra, S.: Probabilistic checking of proofs: a new characterization of NP. J. ACM 45(1), 70–122 (1998). Preliminary version in Proc. of 33rd FOCS, 1992
23. Babai, L., Nisan, N., Szegedy, M.: Multiparty protocols and logspace-hard pseudorandom sequences. In: Proc. of 21st STOC, pp. 1–11 (1989)
24. Barrington, D.A.: Bounded-width polynomial-size branching programs recognize exactly those languages in NC^1. In: Proc. of 18th STOC, pp. 1–5 (1986)
25. Beaver, D., Micali, S., Rogaway, P.: The round complexity of secure protocols (extended abstract). In: Proc. of 22nd STOC, pp. 503–513 (1990)
26. Bellare, M., Yung, M.: Certifying permutations: noninteractive zero-knowledge based on any trapdoor permutation. J. Cryptol. 9(3), 149–166 (1996). Preliminary version in Proc. of CRYPTO '92
27. Ben-Or, M., Goldwasser, S., Wigderson, A.: Completeness theorems for non-cryptographic fault-tolerant distributed computation. In: Proc. of 20th STOC, pp. 1–10 (1988)
28. Ben-Sasson, E., Sudan, M., Vadhan, S., Wigderson, A.: Randomness-efficient low-degree tests and short PCPs via epsilon-biased sets. In: Proc. of 35th STOC, pp. 612–621 (2003)
29. Bennett, C., Grinstein, C.: Role of irreversibility in stabilizing complex and nonenergodic behavior in locally interacting discrete systems. Phys. Rev. Lett. 55, 657–660 (1985)
30. Berlekamp, E., Conway, J., Guy, R.: Winning Ways for Your Mathematical Plays. Academic Press, New York (1983)
31. Berlekamp, E.R., McEliece, R.J., van Tilborg, H.C.: On the inherent intractability of certain coding problems. IEEE Trans. Inf. Theory 24(3), 384–386 (1978)
32. Blum, A., Furst, M., Kearns, M., Lipton, R.J.: Cryptographic primitives based on hard learning problems. In: Advances in Cryptology: Proc. of CRYPTO '93. LNCS, vol. 773, pp. 278–291 (1994)
33. Blum, A., Kalai, A., Wasserman, H.: Noise-tolerant learning, the parity problem, and the statistical query model. J. ACM 50(4), 506–519 (2003). Preliminary version in Proc. of 32nd STOC, 2000
34. Blum, M.: Coin flipping by telephone: a protocol for solving impossible problems. SIGACT News 15(1), 23–27 (1983)
35. Blum, M., Goldwasser, S.: An efficient probabilistic public-key encryption scheme which hides all partial information. In: Advances in Cryptology: Proc. of CRYPTO '84. LNCS, vol. 196, pp. 289–302 (1985)
36. Blum, M., Micali, S.: How to generate cryptographically strong sequences of pseudo-random bits. SIAM J. Comput. 13, 850–864 (1984). Preliminary version in FOCS 82
37. Bogdanov, A., Guo, S.: Sparse extractor families for all the entropy. In: Proc. of 4th ITCS (2012)
38. Bogdanov, A., Qiao, Y.: On the security of Goldreich's one-way function. In: Proc. of 13th RANDOM, pp. 392–405 (2009)
39. Bogdanov, A., Rosen, A.: Input locality and hardness amplification. In: Proc. of 8th TCC, pp. 1–18 (2011)
40. Boppana, R.B., Håstad, J., Zachos, S.: Does co-NP have short interactive proofs? Inf. Process. Lett. 25, 127–132 (1987)
41. Canetti, R.: Security and composition of multiparty cryptographic protocols. J. Cryptol. 13(1), 143–202 (2000)

42. Canetti, R.: Universally composable security: a new paradigm for cryptographic protocols. In: Proc. of 42nd FOCS, pp. 136–145 (2001)

43. Canetti, R., Krawczyk, H., Nielsen, J.: Relaxing chosen ciphertext security of encryption schemes. In: Advances in Cryptology: Proc. of CRYPTO '03. LNCS, vol. 2729, pp. 565–582 (2003)

44. Capalbo, M., Reingold, O., Vadhan, S., Wigderson, A.: Randomness conductors and constant-degree lossless expanders. In: Proc. of 34th STOC, pp. 659–668 (2002)

45. Chaum, D., Crépeau, C., Damgård, I.: Multiparty unconditionally secure protocols (extended abstract). In: Proc. of 20th STOC, pp. 11–19 (1988)

46. Chor, B., Goldreich, O.: Unbiased bits from sources of weak randomness and probabilistic communication complexity. SIAM J. Comput. **17**(2), 230–261 (1988)

47. Cook, J., Etesami, O., Miller, R., Trevisan, L.: Goldreich's one-way function candidate and myopic backtracking algorithms. In: Proc. of 6th TCC, pp. 521–538 (2009). Full version in ECCC TR12-175

48. Cook, S.A.: The complexity of theorem-proving procedures. In: Proc. of 3rd STOC, pp. 151–158. ACM, New York (1971)

49. Cramer, R., Fehr, S., Ishai, Y., Kushilevitz, E.: Efficient multi-party computation over rings. In: Advances in Cryptology: Proc. of EUROCRYPT '03, pp. 596–613 (2003)

50. Cryan, M., Miltersen, P.B.: On pseudorandom generators in NC^0. In: Proc. of 26th MFCS (2001)

51. Damgård, I.B.: Collision free hash functions and public key signature schemes. In: Proc. of Eurocrypt'87, pp. 203–216 (1988)

52. Damgård, I., Ishai, Y.: Scalable secure multiparty computation. In: Advances in Cryptology: Proc. of CRYPTO '06, pp. 501–520 (2006)

53. Damgård, I.B., Pedersen, T.P., Pfitzmann, B.: On the existence of statistically hiding bit commitment schemes and fail-stop signatures. In: Advances in Cryptology: Proc. of CRYPTO '93. LNCS, vol. 773, pp. 250–265 (1994)

54. De, A., Watson, T.: Extractors and lower bounds for locally samplable sources. In: Proc. of 15th RANDOM (2011)

55. Dinur, I., Finucane, H.: Cryptography with locality two. Unpublished manuscript (2011)

56. Dodis, Y., Smith, A.: Correcting errors without leaking partial information. In: Proc. of 37th STOC, pp. 654–663 (2005)

57. Dolev, D., Dwork, C., Naor, M.: Nonmalleable cryptography. SIAM J. Comput. **30**(2), 391–437 (2000)

58. Feige, U.: Relations between average case complexity and approximation complexity. In: Proc. of 34th STOC, pp. 534–543 (2002)

59. Feige, U., Lapidot, D., Shamir, A.: Multiple noninteractive zero knowledge proofs under general assumptions. SIAM J. Comput. **29**(1), 1–28 (2000). Preliminary version in Proc. of 31st FOCS, 1990

60. Feigenbaum, J.: Locally random reductions in interactive complexity theory. In: Advances in Computational Complexity Theory. DIMACS Series on Discrete Mathematics and Theoretical Computer Science, vol. 13, pp. 73–98 (1993)

61. Feistel, H.: Cryptography and computer privacy. Sci. Am. **228**(5), 15–23 (1973)

62. Feldman, V., Gopalan, P., Khot, S., Ponnuswami, A.K.: New results for learning noisy parities and halfspaces. In: Proc. of 47th FOCS, pp. 563–574 (2006)

63. Fortnow, L.: The complexity of perfect zero-knowledge (extended abstract). In: Proc. of 19th STOC, pp. 204–209 (1987)

64. Gamal, T.E.: A public key cryptosystem and a signature scheme based on discrete logarithms. In: Advances in Cryptology: Proc. of CRYPTO '84. LNCS, vol. 196, pp. 10–18 (1985). Also published in IEEE Transactions on Information Theory **IT-31**(4) (1985)

65. Ganguly, N., Sikdar, B.K., Deutsch, A., Canright, G., Chaudhuri, P.P.: A survey on cellular automata. Technical Report 30, Centre for High Performance Computing, Dresden University of Technology (2003). www.cs.unibo.it/bison/publications/CAsurvey.pdf.

66. Goldberg, A.V., Kharitonov, M., Yung, M.: Lower bounds for pseudorandom number generators. In: Proc. of 30th FOCS, pp. 242–247 (1989)
67. Goldreich, O.: A note on computational indistinguishability. Inf. Process. Lett. **34**(6), 277–281 (1990)
68. Goldreich, O.: Modern Cryptography, Probabilistic Proofs and Pseudorandomness. Algorithms and Combinatorics, vol. 17. Springer, Berlin (1998)
69. Goldreich, O.: Candidate one-way functions based on expander graphs. Electron. Colloq. Comput. Complex. **7**, 090 (2000)
70. Goldreich, O.: Foundations of Cryptography: Basic Tools. Cambridge University Press, Cambridge (2001)
71. Goldreich, O.: Randomized methods in computation—lecture notes (2001). http://www.wisdom.weizmann.ac.il/~oded/rnd.html
72. Goldreich, O.: Foundations of Cryptography: Basic Applications. Cambridge University Press, Cambridge (2004)
73. Goldreich, O., Goldwasser, S., Halevi, S.: Collision-free hashing from lattice problems. Electron. Colloq. Comput. Complex. **96**, 042 (1996)
74. Goldreich, O., Goldwasser, S., Micali, S.: How to construct random functions. J. ACM **33**, 792–807 (1986)
75. Goldreich, O., Kahan, A.: How to construct constant-round zero-knowledge proof systems for NP. J. Cryptol. **9**(2), 167–189 (1996)
76. Goldreich, O., Krawczyk, H., Luby, M.: On the existence of pseudorandom generators. SIAM J. Comput. **22**(6), 1163–1175 (1993). Preliminary version in Proc. of 29th FOCS, 1988
77. Goldreich, O., Levin, L.: A hard-core predicate for all one-way functions. In: Proc. of 21st STOC, pp. 25–32 (1989)
78. Goldreich, O., Micali, S., Wigderson, A.: How to play any mental game (extended abstract). In: Proc. of 19th STOC, pp. 218–229 (1987)
79. Goldreich, O., Wigderson, A.: Tiny families of functions with random properties: a quality-size trade-off for hashing. Random Struct. Algorithms **11**(4), 315–343 (1997)
80. Goldwasser, S., Micali, S.: Probabilistic encryption. J. Comput. Syst. Sci. **28**(2), 270–299 (1984). Preliminary version in Proc. of STOC '82
81. Goldwasser, S., Micali, S., Rackoff, C.: The knowledge complexity of interactive proof systems. SIAM J. Comput. **18**(1), 186–208 (1989). Preliminary version in STOC 1985
82. Haitner, I., Harnik, D., Reingold, O.: On the power of the randomized iterate. In: Advances in Cryptology: Proc. of CRYPTO '06, pp. 22–40 (2006)
83. Haitner, I., Reingold, O., Vadhan, S.P.: Efficiency improvements in constructing pseudorandom generators from one-way functions. In: Proc. of 42nd STOC, pp. 437–446 (2010)
84. Halevi, S., Micali, S.: Practical and provably-secure commitment schemes from collision-free hashing. In: Advances in Cryptology: Proc. of CRYPTO '96. LNCS, vol. 1109, pp. 201–215 (1996)
85. Håstad, J.: One-way permutations in NC^0. Inf. Process. Lett. **26**, 153–155 (1987)
86. Håstad, J., Impagliazzo, R., Levin, L.A., Luby, M.: A pseudorandom generator from any one-way function. SIAM J. Comput. **28**(4), 1364–1396 (1999)
87. Holenstein, T.: Pseudorandom generators from one-way functions: a simple construction for any hardness. In: Proc. of 3rd TCC, pp. 443–461 (2006)
88. Hopper, N.J., Blum, M.: Secure human identification protocols. In: Advances in Cryptology: Proc. of ASIACRYPT '01. LNCS, vol. 2248, pp. 52–66 (2001)
89. Hsiao, C.Y., Reyzin, L.: Finding collisions on a public road, or do secure hash functions need secret coins? In: Advances in Cryptology: Proc. of CRYPTO '04. LNCS, vol. 3152, pp. 92–105 (2004)
90. Impagliazzo, R., Luby, M.: One-way functions are essential for complexity based cryptography. In: Proc. of 30th FOCS, pp. 230–235 (1989)
91. Impagliazzo, R., Naor, M.: Efficient cryptographic schemes provably as secure as subset sum. J. Cryptol. **9**, 199–216 (1996)

92. Ishai, Y., Kushilevitz, E.: Randomizing polynomials: a new representation with applications to round-efficient secure computation. In: Proc. of 41st FOCS, pp. 294–304 (2000)

93. Ishai, Y., Kushilevitz, E.: Perfect constant-round secure computation via perfect randomizing polynomials. In: Proc. of 29th ICALP, pp. 244–256 (2002)

94. Ishai, Y., Kushilevitz, E., Li, X., Ostrovsky, R., Prabhakaran, M., Sahai, A., Zuckerman, D.: Robust Pseudorandom Generators. Personal communication (2012)

95. Ishai, Y., Kushilevitz, E., Ostrovsky, R., Sahai, A.: Cryptography with constant computational overhead. In: Proc. of 40th STOC, pp. 433–442 (2008)

96. Janwa, H., Moreno, O.: McEliece public key cryptosystems using algebraic-geometric codes. Des. Codes Cryptogr. 8(3), 293–307 (1996)

97. Juels, A., Weis, S.: Authenticating pervasive devices with human protocols. In: Advances in Cryptology: Proc. of CRYPTO '05, pp. 293–308 (2005)

98. Kam, J.B., Davida, G.I.: Structured design of substitution-permutation encryption networks. IEEE Trans. Comput. 28(10), 747–753 (1979)

99. Katz, J., Shin, J.-S.: Parallel and concurrent security of the HB and HB+ protocols. In: Advances in Cryptology: Proc. of Eurocrypt'06, pp. 73–87 (2006)

100. Katz, J., Yung, M.: Complete characterization of security notions for probabilistic private-key encryption. In: Proc. of 32nd STOC, pp. 245–254 (2000)

101. Kearns, M., Mansour, Y., Ron, D., Rubinfeld, R., Schapire, R.E., Sellie, L.: On the learnability of discrete distributions. In: Proc. of 26th STOC, pp. 273–282 (1994)

102. Kearns, M.J.: Efficient noise-tolerant learning from statistical queries. J. ACM 45(6), 983–1006 (1998)

103. Kharitonov, M.: Cryptographic hardness of distribution-specific learning. In: Proc. of 25th STOC, pp. 372–381 (1993)

104. Kilian, J.: Founding cryptography on oblivious transfer. In: Proc. of 20th STOC, pp. 20–31 (1988)

105. Krause, M., Lucks, S.: On the minimal hardware complexity of pseudorandom function generators (extended abstract). In: Proc. of 18th STACS. LNCS, vol. 2010, pp. 419–430 (2001)

106. Levin, L.A.: Universal sequential search problems. PINFTRANS: Problems of Information Transmission (translated from Problemy Peredachi Informatsii (Russian)) 9 (1973)

107. Lindell, Y., Pinkas, B.: A proof of Yao's protocol for secure two-party computation. Electron. Colloq. Comput. Complex. 11, 063 (2004)

108. Linial, N., Mansour, Y., Nisan, N.: Constant depth circuits, Fourier transform, and learnability. J. ACM 40(3), 607–620 (1993). Preliminary version in Proc. of 30th FOCS, 1989

109. Lyubashevsky, V.: The parity problem in the presence of noise, decoding random linear codes, and the subset sum problem. In: Proc. of 9th RANDOM (2005)

110. McEliece, R.J.: A public-key cryptosystem based on algebraic coding theory. Technical Report DSN PR 42-44, Jet Prop. Lab. (1978)

111. Moore, E.F. (ed.): Sequential Machines: Selected Papers. Addison-Wesley/Longman, Reading/Harlow (1964)

112. Mossel, E., Shpilka, A., Trevisan, L.: On ε-biased generators in NC^0. In: Proc. of 44th FOCS, pp. 136–145 (2003)

113. Naor, J., Naor, M.: Small-bias probability spaces: efficient constructions and applications. SIAM J. Comput. 22(4), 838–856 (1993). Preliminary version in Proc. of 22nd STOC, 1990

114. Naor, M.: Bit commitment using pseudorandomness. J. Cryptol. 4, 151–158 (1991)

115. Naor, M., Pinkas, B., Sumner, R.: Privacy preserving auctions and mechanism design. In: Proc. of 1st ACM Conference on Electronic Commerce, pp. 129–139 (1999)

116. Naor, M., Reingold, O.: Synthesizers and their application to the parallel construction of pseudo-random functions. J. Comput. Syst. Sci. Int. 58(2), 336–375 (1999)

117. Naor, M., Reingold, O.: Number-theoretic constructions of efficient pseudo-random functions. J. ACM 51(2), 231–262 (2004). Preliminary version in Proc. of 38th FOCS, 1997

118. Neumann, J.V.: Theory of Self-Reproducing Automata. University of Illinois Press, Champaign (1966)
119. Nisan, N.: Pseudorandom generators for space-bounded computation. Combinatorica **12**(4), 449–461 (1992)
120. Panjwani, S.K.: An experimental evaluation of Goldreich's one-way function. Technical report, IIT, Bombay (2001). http://www.wisdom.weizmann.ac.il/~oded/PS/ow-report.ps
121. Papadimitriou, C.H., Yannakakis, M.: Optimization, approximation, and complexity classes. J. Comput. Syst. Sci. Int. **43**, 425–440 (1991). Preliminary version in Proc. of 20th STOC, 1988
122. Pedersen, T.: Non-interactive and information-theoretic secure verifiable secret sharing. In: Advances in Cryptology: Proc. of CRYPTO '91. LNCS, vol. 576, pp. 129–149 (1991)
123. Rabin, M.O.: Digitalized signatures and public key functions as intractable as factoring. Technical Report 212, LCS, MIT (1979)
124. Radhakrishnan, J., Ta-Shma, A.: Tight bounds for depth-two superconcentrators. SIAM J. Discrete Math. **13**(1), 2–24 (2000). Preliminary version in Proc. of 38th FOCS, 1997
125. Regev, O.: New lattice based cryptographic constructions. In: Proc. of 35th STOC, pp. 407–416 (2003)
126. Reingold, O., Trevisan, L., Vadhan, S.: Notions of reducibility between cryptographic primitives. In: Proc. of 1st TCC '04. LNCS, vol. 2951, pp. 1–20 (2004)
127. Rivest, R.L., Shamir, A., Adleman, L.M.: A method for obtaining digital signatures and public-key cryptosystems. Commun. ACM **21**(2), 120–126 (1978)
128. Rogaway, P.: The round complexity of secure protocols. PhD thesis, MIT (1991)
129. Sahai, A., Vadhan, S.: A complete problem for statistical zero knowledge. J. ACM **50**(2), 196–249 (2003). Preliminary version in FOCS 1997
130. Shannon, C.E.: Communication theory of secrecy systems. Bell Syst. Tech. J. **28**(4), 656–715 (1949)
131. Shpilka, A.: Constructions of low-degree and error-correcting ε-biased generators. In: Proc. of 21st Conference on Computational Complexity (CCC), pp. 33–45 (2006)
132. Sudan, M.: Algorithmic Introduction to Coding Theory—Lecture Notes. http://theory.csail. mit.edu/~madhu/FT01/ (2002)
133. Ta-Shma, A., Umans, C., Zuckerman, D.: Loss-less condensers, unbalanced expanders, and extractors. In: Proc. of 33rd STOC, pp. 143–152 (2001)
134. Tate, S.R., Xu, K.: On garbled circuits and constant round secure function evaluation. CoPS Lab Technical Report 2003-02, University of North Texas (2003)
135. Varshamov, R.R.: Estimate of the number of signals in error correcting codes. Dokl. Akad. Nauk SSSR **117**, 739–741 (1957)
136. Vichniac, G.Y.: Simulating physics with cellular automata. Phys. D, Nonlinear Phenom. **10**(1–2), 96–115 (1984)
137. Viola, E.: The complexity of constructing pseudorandom generators from hard functions. Comput. Complex. **13**(3–4), 147–188 (2005)
138. Viola, E.: On constructing parallel pseudorandom generators from one-way functions. In: Proc. of 20th Conference on Computational Complexity (CCC), pp. 183–197 (2005)
139. Viola, E.: Extractors for circuit sources. In: Proc. of 52nd FOCS, pp. 220–229 (2011)
140. Viola, E.: The complexity of distributions. SIAM J. Comput. **41**(1), 191–218 (2012)
141. Webster, A.F., Tavares, S.E.: On the design of s-boxes. In: Advances in Cryptology: Proc. of CRYPTO '85, pp. 523–534. Springer, Berlin (1986)
142. Wigderson, A.: NL/*poly* ⊆ ⊕L/*poly*. In: Proc. of 9th Structure in Complexity Theory Conference, pp. 59–62 (1994)
143. Wolfram, S.: Universality and complexity in cellular automata. Phys. D, Nonlinear Phenom. **10**(1–2), 1–35 (1984)
144. Yao, A.C.: Theory and application of trapdoor functions. In: Proc. of 23rd FOCS, pp. 80–91 (1982)

145. Yao, A.C.: How to generate and exchange secrets. In: Proc. of 27th FOCS, pp. 162–167 (1986)
146. Yap, C.-K.: Some consequences of non-uniform conditions on uniform classes. Theor. Comput. Sci. **26**, 287–300 (1983)
147. Yu, X., Yung, M.: Space lower-bounds for pseudorandom-generators. In: Proc. of 9th Structure in Complexity Theory Conference, pp. 186–197 (1994)

Printed in the United States
By Bookmasters